Energy Production and Management
in the 21st Century V

WITPRESS

WIT Press publishes leading books in Science and Technology.
Visit our website for the current list of titles.
www.witpress.com

WITeLibrary

Home of the Transactions of the Wessex Institute.
Papers contained in this volume are archived in the WIT eLibrary in volume 255 of WIT
Transactions on Ecology and the Environment (ISSN 1743-3541).
The WIT eLibrary provides the international scientific community with immediate and
permanent access to individual papers presented at WIT conferences.
Visit the WIT eLibrary at www.witpress.com.

FIFTH INTERNATIONAL CONFERENCE ON
ENERGY PRODUCTION AND MANAGEMENT: THE
QUEST FOR SUSTAINABLE ENERGY

Energy Production and Management 2022

CONFERENCE CHAIRS

E. Magaril
Ural Federal University, Russia

S. Syngellakis
Wessex Institute, UK
Member of the WIT Board of Directors

INTERNATIONAL SCIENTIFIC ADVISORY COMMITTEE

Organised by
Wessex Institute, UK

Sponsored by
WIT Transactions on Ecology and the Environment
International Journal of Energy Production and Management

WIT Transactions

Wessex Institute
Ashurst Lodge, Ashurst
Southampton SO40 7AA, UK

We would like to express thanks to all the conference Chairs and members of the International Scientific Advisory Committee for their efforts during the 2022 conference season.

Conference Chairs

Joanna Barnes
University of the West of England, UK

Juan Casares
University of Santiago de Compostela, Spain
(Member of WIT Board of Directors)

Alexander Cheng
University of Mississippi, USA
(Member of WIT Board of Directors)

Pilar Chias
University of Alcala, Spain

Pablo Diaz Rodriguez
University of La Laguna, Spain

Andrea Fabbri
University of Milano-Bicocca, Italy

Fabio Garzia
University of Rome "La Sapienza", Italy

Massimo Guarascio
University of Rome "La Sapienza", Italy

Santiago Hernandez
University of A Coruna, Spain
(Member of WIT Board of Directors)

Massimiliano Lega
University of Naples Parthenope, Italy

Mara Lombardi
University of Rome "La Sapienza", Italy

James Longhurst
University of the West of England, UK

Elena Magaril
Ural Federal University, Russia

Stefano Mambretti
Polytechnic of Milan, Italy
(Member of WIT Board of Directors)

Jose Manuel Mera
Polytechnic University of Madrid, Spain

Jose Luis Miralles i Garcia
Polytechnic University of Valencia, Spain

Giorgio Passerini
Polytechnic University of Le Marche, Italy
(Member of WIT Board of Directors)

David Proverbs
University of Wolverhampton, UK

Elena Rada
Insubria University, Italy

Stefano Ricci
University of Rome, La Sapienza

Graham Schleyer
University of Liverpool, UK

Stavros Syngellakis
Wessex Institute, UK
(Member of WIT Board of Directors)

International Scientific Advisory Committee Members 2022

Borna Abramovic University of Zagreb, Croatia

Tawfiq Abuhantash American University of Ras Al Khaimah, UAE

Alejandro Acosta Collazo Autonomous University of Aguascalientes, Mexico

Khalid Al Saud King Saud University, Saudi Arabia

Ghassan Al-Dweik Palestine Polytechnic University, Palestine

Hind Algahtani Imam Abdulrahman bin Faisal University, Saudi Arabia

Abdulkader Algilani King Abdulaziz University, Saudi Arabia

Mir Ali University of Illinois at Urbana-Champaign, USA

Bakari Aliyu Taraba State University, Nigeria

Samar Aljahdali University of Jeddah, Saudi Arabia

Hussain Al-Kayiem Universiti Teknologi PETRONAS, Malaysia

Jose Ignacio Alonso Polytechnic University of Madrid, Spain

Reem Alsabban University of Jeddah, Saudi Arabia

Sultan Al-Salem KISR, Kuwait

Andrea Antonucci University of Roma 3, Italy

Srazali Aripin International Islamic University Malaysia, Malaysia

Eman Assi American University of Ras Al Khaimah, UAE

Sahar Attia Cairo University, Egypt

Jihad Awad Ajman University, UAE

Warren Axelrod C. Warren Axelrod LLC, USA

Mohammed Bagader Umm Al-Qura University, Saudi Arabia

Azizi Bahauddin Universiti Sains Malaysia, Malaysia

Francine Baker Wolfson College, UK

Marco Baldi Marche Polytechnic University, Italy

Michael Barber University of Utah, USA

Socrates Basbas Aristotle University of Thessaloniki, Greece

Joao Batista University of São Paulo, Brazil

Gianfranco Becciu Politecnico di Milano, Italy

Michael Beer Leibniz Universitat Hannover, Germany

Khadija Benis c5Lab, Portugal

Marco Bietresato Free University of Bolzano, Italy

Alma Bojorquez-Vargas Autonomous University of San Luis Potosí, Mexico

Daniel Bonotto UNESP, Brazil

Colin Booth University of the West of England, UK

Carlos Borrego University of Aveiro, Portugal

Bouzid Boudiaf Ajman University, UAE

Zuzana Boukalova VODNÍ ZDROJE, a.s., Czech Republic

Djamel Boussaa Qatar University, Qatar

Roman Brandtweiner Vienna University of Economics and Business, Austria

Roger Brewster Bond University, Australia

Andre Buchau University of Stuttgart, Germany

Raul Campos RCQ Structural Engineering, Chile

Richard Carranza Carranza Consulting, USA

Paul Carrion Mero ESPOL Polytechnic University, Ecuador

Joao-Manuel Carvalho Universidade de Lisboa, Portugal

Ana Cristina Paixao Casaca Isel, Portugal

Ricardo Castedo Universidad Politécnica de Madrid, Spain

Robert Cerny Czech Technical University Prague, Czech Republic

Camilo Cerro American University of Sharjah, UAE

Vicent Esteban Chapapria Polytechnic University of Valencia, Spain

Galina Chebotareva Ural Federal University, Russia

Hai-Bo Chen University of Science & Technology of China, China

Jeng-Tzong Chen National Taiwan Ocean University, Taiwan

Weiqiu Chen Zhejiang University, China

Rémy Chevrier SNCF Innovation & Research, France

Energy Production and Management in the 21st Century V

The Quest for Sustainable Energy

Editors

S. Syngellakis
Wessex Institute, UK
Member of the WIT Board of Directors

E. Magaril
Ural Federal University, Russia

WITPRESS Southampton, Boston

Editors:

S. Syngellakis
Wessex Institute, UK
Member of the WIT Board of Directors

E. Magaril
Ural Federal University, Russia

Published by

WIT Press
Ashurst Lodge, Ashurst, Southampton, SO40 7AA, UK
Tel: 44 (0) 238 029 3223; Fax: 44 (0) 238 029 2853
E-Mail: witpress@witpress.com
http://www.witpress.com

For USA, Canada and Mexico

Computational Mechanics International Inc
25 Bridge Street, Billerica, MA 01821, USA
Tel: 978 667 5841; Fax: 978 667 7582
E-Mail: infousa@witpress.com
http://www.witpress.com

British Library Cataloguing-in-Publication Data

A Catalogue record for this book is available
from the British Library

ISBN: 978-1-78466-457-2
eISBN: 978-1-78466-458-9
ISSN: 1746-448X (print)
ISSN: 1743-3541 (on-line)

*The texts of the papers in this volume were set individually by the authors or under their
supervision. Only minor corrections to the text may have been carried out by the publisher.*

Preface

This book contains a selection of papers among those presented at the 5th International Conference on Energy Production and Management: The Quest for Sustainable Energy, organized by the Wessex Institute of Technology, UK. The meeting was sponsored by the WIT Transactions on Ecology and the Environment and the International Journal of Energy Production and Management.

The aim of the Conference was to provide a forum for discussion on issues concerning the future of energy production, operation and management in a changing world. The energy sector faces major challenges and complex problems that need to be addressed towards pursuing economic and social development.

In the current technogenic environment, it is very important to introduce innovative processes and technologies which enable the improvement of exploitation of conventional fuels, thus increasing energy efficiency in power generation, industry and transportation as well as reducing environmental pollution. A contribution to these objectives is the development of predictive models for the lower limit of coupling physical properties of shale oil reservoirs for the application of CO_2 huff-n-puff technology under different economic and technological development conditions.

The complexity of modern energy production and management requires a multidisciplinary approach that can take into consideration not only advances in technology but also involves economic, social, political and environmental, aspects. High economic risks, for instance, threaten the sustainable operation of energy systems. Issues related to reliability and survivability of energy supply systems can be studied taking into account the structural systematic characteristics of power cogeneration systems and conditions contributing to appearance of system accidents leading to the loss of sustainability. In the electric power industry, economic benefits can be achieved from the application of platform tools namely, knowledge-intensive services based on integration of utilities, consumers, developers and digital solutions providers into single eco-systems. Normative documents are adopted in various countries to assess the environmental and economic efficiency of investment projects. Key differences in regulations and the main problems and difficulties of conducting such assessments can be identified by critically reviewing these documents.

The massive consumption of fossil fuels due to the rapid growth of the world population and the demand for higher living standards has had serious environmental consequences. Hence, there is an urgent need to increase the use of renewable energy (RE) sources. This transition from an economy based on conventional fuels to one relying on renewable sources requires, apart from consideration of national priorities and opportunities, new management concepts and tools. Using

RE resources for hybrid power generation involves uncertainty and variability owing to RE features. Demand-side response (DR) combined with artificial Intelligence techniques is proposed as a means of estimating the availability of generated RE power thus providing the electricity system with demand flexibility.

Biomass is a renewable source due to its potential for thermochemical transformation into biofuel. Biogas projects at wastewater treatment plants reduce the negative environmental waste impact and contribute to independence from fossil fuels. It is important that financial instruments for implementing such projects are available to a developing country and barriers for their development identified.

The Water-Energy-Food (WEF) nexus is a new development strategy that reduces the carbon footprint, provides energy and food security while maintaining the relationship between socio-economic progress and environmental protection. A study exploring the growth of this academic field using bibliometric review models, shows that the WEF nexus approach to energy developments creates new prospects for decision-making for socio-economic, political, and environmental progress. By applying tools that address the integrated WEF nexus, systems can be considered holistically, especially in the planning and production stages. A systematic selection approach can be adopted whereby multi-criteria decision making is applied to determine which tool best fits a set of circumstances.

Very significant reductions in water and water-related energy demands and associated emissions and costs are possible in leisure centres without impacting service quality and delivery. This can be achieved through a water management hierarchy which prioritises management actions in order of preference of implementation; such interventions must be considered on a site-specific and case-by-case basis.

Energy savings in the built environment attract particular attention due to their beneficial effect on sustainability. Community engagement with the energy efficiency in housing, explored via semi-structured, in-depth interviews with stakeholders, provides a sense about interactions among actors involved in the renovation of owner-occupied single-family houses and the role of both social and professional networks within communities in promoting energy policies enlargement and effectiveness. A predictive maintenance strategy, based on machine learning systems, contributes to the efficiency of residential mechanical and electrical plant systems by drastically reducing their malfunctioning thus leading to substantial improvements in their overall operation. A proposed dedicated household tracking system also uses machine learning technologies to predict and manipulate the energy consumption of household appliances and then apply the most suitable strategy for energy conservation.

This volume is part of the WIT Transactions on Ecology and the Environment. The digital version of the papers, as well as those presented in the previous conferences held in Yekaterinburg, Russia in 2014, Ancona, Italy in 2016, the New Forest, UK in 2018 and online in 2020 are archived in Open Access format in the eLibrary of the Wessex Institute (https://www.witpress.com/elibrary) where they are freely available to the international community.

The editors are grateful to all authors for the quality of their contributions as well as to the members of the International Scientific Advisory Committee and other colleagues who helped to review the papers and hence ensure the quality of this volume.

The Editors, 2022

Contents

Section 3: Energy and the built environment

SECTION 1
ENERGY PRODUCTION
AND ECONOMICS

LOWER LIMITS OF COUPLING PHYSICAL PROPERTIES OF SHALE OIL RESERVOIRS FOR THE APPLICATION OF CO$_2$ HUFF-N-PUFF

PENG WANG, SHIJUN HUANG & FENGLAN ZHAO
China University of Petroleum (Beijing), China

ABSTRACT

CO$_2$ flooding for enhanced shale oil recovery is unsatisfactory in pilot projects due to the fracture network complexity induced by gas channelling. CO$_2$ huff-n-puff has been proven to improve the estimated ultimate recovery (EUR) for a single well. The purpose of this work is to determine the lower limit of coupling physical properties of shale oil reservoirs for the application of CO$_2$ huff-n-puff technology under different economic and technological development conditions. In this work, a numerical model with two horizontal wells is established to simulate CO$_2$ huff-n-puff process in shale oil reservoirs. Logarithmically spaced, locally refined, and dual permeability (LS-LR-DK) model is used to generate hydraulic fractures. Then, the critical shifts of hydrocarbon molecules confined in shale nanopores are corrected by a modified Soave–Redlich–Kwong equation of state (m-SRK EOS) to model the phase behaviour accurately. Subsequently, numerical simulations are conducted to investigate the influence of horizontal well length, huff-n-puff cycle, and matrix permeability on well productivity. Finally, the lower limits of coupling the permeability and initial oil saturation are determined in the economic limit output. Longer horizontal well shows a much better performance on improving single-well EUR both in the depletion and CO$_2$ huff-n-puff stage. There is an optimum injection time for CO$_2$ huff-n-puff under different reservoir conditions. Determining the lower limit of coupling physical properties in shale oil reservoirs is critical from the perspective of investment income. When the matrix permeability is in the range of 0.01mD to 0.1mD, the performance on improving single-well EUR of CO$_2$ huff-n-puff is better, and it increases as the initial oil saturation increases.
Keywords: shale oil reservoirs, CO$_2$ huff-n-puff, lower limits of coupling physical properties, nano-confinement, numerical simulation.

1 INTRODUCTION

With a significant increase in the global energy demand and a gradually declined production of conventional oil reservoirs, unconventional resource has become the focus of researchers [1]. Several studies have demonstrated that the estimated recoverable reserves of shale oil are extremely large, and it has become one of the most potential and important alternative resource [2], [3]. Shale reservoirs have extremely low porosity and permeability resulting from the dominance of nano-scale pores and the heterogeneity of pore structures [4], [5]. Hence, conventional technologies are no longer applicable in shale oil reservoirs. The advancements of horizontal wells and hydraulic fracturing technology have promoted the development of shale reservoirs [6]. There is a high oil production rate during the early age of depletion, but it declines sharply in the succeeding years, and the final oil recovery factor of depletion is generally less than 7% [7]. The main reason is that the supplement of the matrix is difficult to satisfy the production from the complex fracture system. Subsequently, infill drilling technology is practiced to improve shale oil production. Although infill drilling can increase the initial oil production, it can only guarantee a relatively high initial production rate [8]. The CO$_2$-EOR method is widely used in conventional oil reservoirs, which draws the attention of the researchers for the application on shale oil reservoirs [9]–[11]. Compared with N$_2$ and CH$_4$, CO$_2$ and shale oil are easier to miscible. Amount of numerical simulation and experimental studies indicate that CO$_2$ injection has high potential for improving shale

WIT Transactions on Ecology and the Environment, Vol 255, © 2022 WIT Press
www.witpress.com, ISSN 1743-3541 (on-line)
doi:10.2495/EPM220011

oil production and playing the role of CO_2 geological storage [12]–[14]. Unfortunately, due to gas channelling induced by fracture network complexity, and the performance of CO_2 flooding in the field the latter is not ideal [15]. Compared with CO_2 flooding, CO_2 huff-n-puff technology has been demonstrated to be feasible in improving single-well estimated ultimate recovery (EUR) [16], [17]. The research of Kong et al. [18] demonstrated that asynchronous CO_2 huff-n-puff performed much better than a synchronous pattern. However, the lower limits for the application of CO_2 huff-n-puff in shale oil reservoirs by considering coupling physical properties is absent.

In this work, numerical simulation methods are utilized to investigate the performance of CO_2 huff-n-puff in stimulated shale oil reservoirs. Compositional simulator based on the conventional EOS is adopted to establish a numerical model representing the rock and fluid properties of Chang 7 shale oil reservoirs in Ordos basin and simulate the composition change during depletion and CO_2 huff-n-puff. Then, a modified SRK EOS proposed in our previous work is utilized to correct the critical temperature and pressure shifts of shale oil components by considering the adsorption induced confinement effect in shale nanopores. Subsequently, numerical simulations are conducted to analyse the sensitiveness of the horizontal well length, huff-n-puff cycle and matrix permeability. Finally, the lower limits of coupling the initial oil saturation and permeability are determined in the economic limit output.

2 METHODOLOGY

A logarithmically spaced, locally refined, and dual permeability (LS-LR-DK) model is used to establish a numerical model to represent the characteristics of Chang 7 shale oil reservoirs in Ordos Basin. The schematic diagram of the numerical reservoir model is shown in Fig. 1. In the LS-LR-DK model, hydraulic fractures are generated through the equivalent method of fracture conductivity, and a Tartan grid is formed through automatic local logarithmic refined. It can efficiently and accurately establish fracture models, improve the computational efficiency, and accurately simulate the characteristics of transient flow in shale oil reservoirs [19].

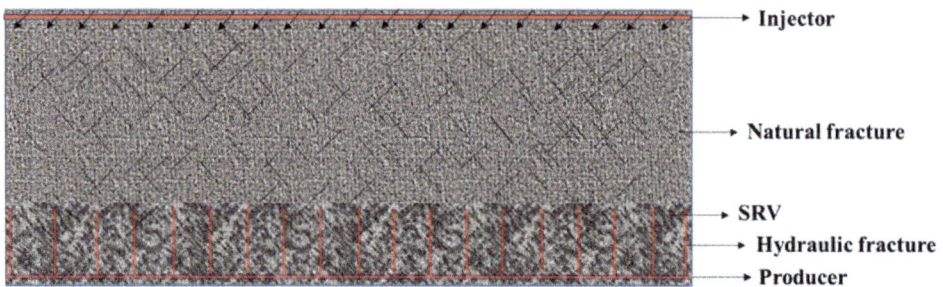

Figure 1: Schematic diagram of the numerical model.

2.1 Numerical model

In this work, a numerical reservoir model is established using the reservoir simulation software developed by Computer Modelling Group (CMG). CMG software has an integrated reservoir simulation framework for unconventional reservoirs to generate the complicated fracture network, forecast future behaviour, and evaluate EOR processes. The dimensions of the numerical reservoir model are 1520 m × 420 m × 12 m, which refer to the length, width and thickness, respectively. A horizontal well is stimulated as a producer, and another horizontal well 400 m away from the producer acts as an injector. In stimulated reservoir volume (SRV), 16 hydraulic fractures are generated, whose fracture spacing and half-length are 100 m and 120 m, respectively. Hydraulic fractures penetrate the numerical reservoir model and their fracture conductivity is 100 mD·m. Other model parameters are provided in Table 1.

Table 1: Other basic parameters of the numerical reservoir model.

Parameter	Value	Parameter	Value
Depth (m)	2,100	Matrix permeability (mD)	0.005
Reservoir temperature (°C)	71.71	Matrix porosity	0.1
Reservoir pressure (kPa)	19,070	Initial oil saturation	0.7

2.2 Fluid model

Under the reservoir conditions, the oil density of Chang 7 oil sample is 722.3 kg/m^3, the gas–oil ratio is 105.05 m^3/m^3, and the saturation pressure is 11,520 kPa. The detailed original oil compositional data is shown in Table 2. To accurately characterize the interaction of CO_2 and oil components and reveal the mechanism of enhanced shale oil recovery of CO_2 huff-n-puff, WINPROP module in CMG software is utilized to establish the fluid model.

Table 2: Original oil compositional data of Chang 7 shale oil sample.

Component	Mol.%	Component	Mol.%
N_2	1.369	iC_5	1.563
CO_2	0.11	nC_5	2.218
CH_4	27.322	C_6	3.515
C_2H_6	8.657	C_7	4.623
C_3H_8	10.051	C_8	3.927
iC_4	1.638	C_9	3.409
nC_4	4.454	C_{10}	2.799
C_{11+}	24.346		

2.2.1 Splitting and lumping

Establishing an accuracy fluid model is a premise for the reliability of a compositional simulation. The compositional simulator is more time-consuming than the black-oil simulator due to more components in original oil composition, but it can more accurately characterize the composition changes during oil production, especially for gas injection. Therefore, it is necessary to split and lump the components for accurately characterizing the fluid phase behaviour and improving the calculation efficiency. Compositional data and binary interaction parameters (BIP) of the fluid model for Chang 7 shale oil sample are listed in Tables 3 and 4.

2.2.2 Modifying the critical property shift

The adsorption induced confinement effect in shale nanopores may alter the fluid thermophysical properties, such as the critical properties. It is crucial to modify the critical

Table 3: Compositional data of fluid model for Chang 7 shale oil sample.

Component	Mol	P_c (atm)	T_c (K)	Acentric factor	MW	SG	Parachor
CO_2	0.0011	72.80	304.20	0.225	44.01	0.82	78.00
N_2-C_1	0.2869	44.86	187.38	0.010	16.61	0.32	75.28
C_2-C_6	0.3210	39.90	395.71	0.171	50.89	0.53	168.45
C_7-C_{10}	0.1476	24.05	576.48	0.410	118.40	0.71	364.42
C_{11}-C_{21}	0.1739	15.41	791.79	0.912	248.62	0.80	559.94
C_{22}-C_{30}	0.0446	9.29	915.44	1.254	393.99	0.86	851.89
C_{31+}	0.0249	6.95	1121.75	1.826	700.71	0.92	1099.14

Table 4: BIP of fluid model for Chang 7 shale oil sample.

Component	CO_2	N_2-C_1	C_2-C_6	C_7-C_{10}	C_{11}-C_{21}	C_{22}-C_{30}	C_{31+}
CO_2	0.000	0.100	0.122	0.115	0.150	0.150	0.150
N_2-C_1	0.100	0.000	0.003	0.013	0.021	0.031	0.041
C_2-C_6	0.122	0.003	0.000	0.003	0.008	0.014	0.022
C_7-C_{10}	0.115	0.013	0.003	0.000	0.001	0.004	0.009
C_{11}-C_{21}	0.150	0.021	0.008	0.001	0.000	0.001	0.004
C_{22}-C_{30}	0.150	0.031	0.014	0.004	0.001	0.000	0.001
C_{31+}	0.150	0.041	0.022	0.009	0.004	0.001	0.000

temperature and pressure shift to model the phase behaviour of nano-confined fluids more accurately. In this section, an m-SRK EOS proposed in our previous work is used to modify the critical property shift of nano-confined fluids; the dimensionless shifts of critical temperature are expressed as following [20]:

$$\Delta T_c = \frac{T_c - T_c'}{T_c} = \frac{\eta(2\psi-1)}{(\psi-1)^2 + \eta(2\psi-1)},$$ (1)

$$\Delta P_c = \frac{P_c - P_c'}{P_c} = \frac{2\eta(2\psi-1)(\psi-1)^2 + [\eta(2\psi-1)]^2}{(\psi-1)^4 + [\eta(2\psi-1)]^2},$$ (2)

where ΔT_c is the critical temperature shift; T_c is the critical temperature, K; T_c' is the critical temperature of nano-confined fluids, K; ΔP_c is the critical pressure shift; P_c is the critical pressure, Pa; P_c' is the critical pressure of nano-confined fluids, Pa; η is relative effective molecular volume coefficient; ψ is fluid distribution coefficient. The detailed process for proposing two analytical formulas of critical property shifts is derived in our previous work. The calculated critical temperature and pressure of nano-confined fluids are shown in Table 5.

After modifying the critical property shift, the calculated values of the fluid model are contrasted with the experimental data of a constant composition expansion (CCE) experiment. As illustrated in Figs 2 and 3, the calculated values of relative volume and oil density for fluid model displays a good agreement compared with the experimental values.

Table 5: Modifying the critical property shift of each oil components.

P_c (atm)	P_c' (atm)	T_c (K)	T_c' (K)
72.80	69.90	304.20	298.08
44.86	43.07	187.38	183.61
39.90	37.85	395.71	385.42
24.05	22.45	576.48	557.00
15.41	14.13	791.79	758.22
9.29	8.36	915.44	868.33
6.95	6.15	1121.75	1054.91

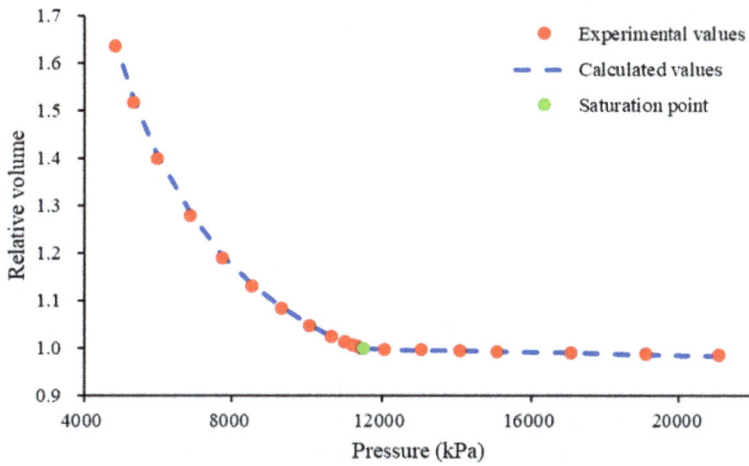

Figure 2: Experimental values and calculated values of relative volume.

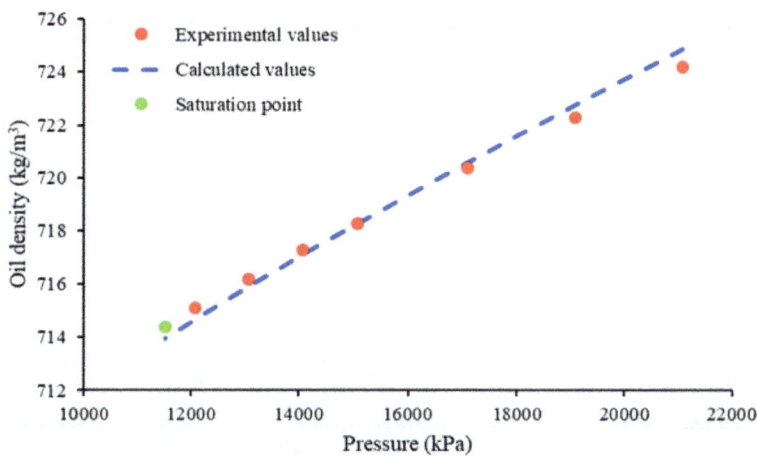

Figure 3: Experimental values and calculated values of oil density.

2.3 CO$_2$ huff-n-puff process

Based on the numerical model and fluid model of Chang 7 shale oil reservoirs, the huff-n-puff process is simulated to investigate the influencing factor and performance for improving single-well EUR. Depletion, as the primary enhanced shale oil recovery method, is conducted first combining it with hydraulic fracturing technology. Then, the producer is shut in and CO$_2$ is injected through another horizontal well. Next, the injector is shut in and the soaking stage goes ahead. Finally, the producer is opened to recover the residual shale oil and the process of injection-soaking-production is repeated for a few cycles.

3 RESULTS AND DISCUSSIONS

In this section, sensitivity analyses of horizontal well length, huff-n-puff cycle, and matrix permeability are performed to investigate the influencing factors on improving Single-well EUR for CO$_2$ huff-n-puff. CO$_2$ is injected after 10 years of depletion, and the CO$_2$ huff-n-puff period lasts 10 years. Then, the lower limits of coupling the permeability and initial oil saturation are determined in the economic limit output.

3.1 Effect of horizontal well length

Fig. 4 shows the oil rate and oil recovery factor of producer with horizontal well lengths of 600 m, 1,000 m and 1,500 m. It can be seen from the Fig. 4 that as the horizontal well length increases, the final oil recovery factor increases significantly. During the depletion, the oil recovery factors of producer with different horizontal well lengths (600 m, 1,000 m and 1,500 m) were 3.64%, 5.40% and 6.86%, while the improved oil recovery factors were 2.70%, 3.46% and 3.75% in CO$_2$ huff-n-puff stage. A longer horizontal well shows a much better performance in shale oil production both at the depletion and CO$_2$ huff-n-puff stage. In such a poor reservoir condition (0.005 mD), the oil recovery factor is difficult to reach 10% during the depletion, and CO$_2$ huff-n-puff can play a good role in improving EUR for a single well. From the perspective of the oil rate, it increases as the horizontal well length decreases after well opening in each huff and puff cycle, but it is more difficult to maintain a high oil rate. This phenomenon is related to re-pressurization in the soaking period the shorter the horizontal well length, the higher the re-pressurization after gas injection. When the horizontal well length equals 1,500 m, the maximum oil rate is gradually increasing in the first five huff-n-puff cycle. Extra-low matrix permeability results in a slow sweep

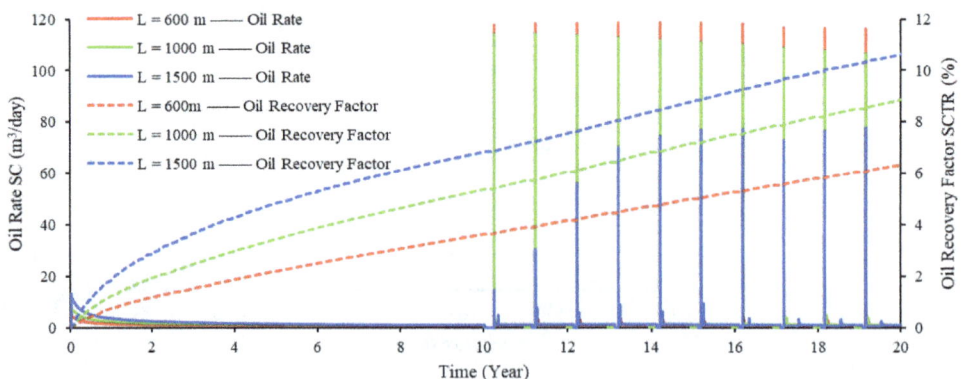

Figure 4: Effect of horizontal well length on the performance of CO$_2$ huff-n-puff.

WIT Transactions on Ecology and the Environment, Vol 255, © 2022 WIT Press
www.witpress.com, ISSN 1743-3541 (on-line)

efficiency of CO_2 injection. With the increase of huff-n-puff cycles, the larger the swept area of CO_2, the better the performance on improving EUR, and the oil rate remains stable in the last five cycles.

3.2 Effect of huff-n-puff cycles

In this section, the influence of huff-n-puff cycles on improving Single-well EUR is explored. Two scenarios of CO_2 huff-n-puff are adopted that are described in Table 6. Fig. 5 reveals the comparison of CO_2 huff-n-puff with ten and two huff-n-puff cycles on improving Single-well EUR. The performance on improving Single-well EUR of ten huff-n-puff cycles is slightly better than that of two huff-n-puff cycles. On the one hand, the total injection volume of CO_2 with two huff-n-puff cycles is less than that of ten huff-n-puff cycles; on the other hand, due to the extra-low matrix permeability, the injected CO_2 is unswept to SRV resulting in a lower oil rate in the first cycle of scenario 1. The main mechanism for improving Single-well EUR is re-pressurization. It suggests that there are differences in the swept efficiency of CO_2 injection under different geological conditions of shale reservoirs, which will affect the scheme formulation of CO_2 huff-n-puff. Therefore, the effect of matrix permeability on improving Single-well EUR will be discussed in the next subsection.

Table 6: Two scenarios of CO_2 huff-n-puff.

Scenario	1	2
Injection cycle	2	10
Injection period (month)	6	2
Injection rate (m^3/day)	100,000	100,000
Soaking period (month)	3	1
Production period (month)	51	9

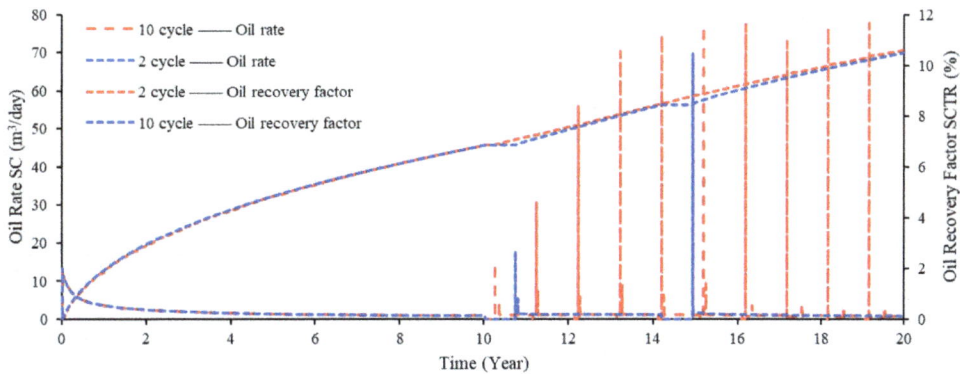

Figure 5: Effect of huff-n-puff cycles on the performance of CO_2 huff-n-puff.

3.3 Effect of matrix permeability

Fig. 6 shows the performance of CO_2 huff-n-puff with different matrix permeability. It can be seen from Fig. 6(a) that the greater the matrix permeability, the greater the maximum oil rate in the initial stage of the depletion, and the shorter the effective huff-n-puff cycle in CO_2

huff-n-puff stage. As depicted in Fig. 6(b), with the increase of matrix permeability, the final oil recovery factor gradually increases. When the matrix permeability is 0.2 and 0.5 mD, the final oil recovery factor is almost identical. In the case of the matrix permeability equalling 0.5 mD, the oil recovery factor remains a constant at around fourth year of the depletion. It indicates that there is an optimized opportunity of CO_2 injection under different reservoir conditions. For the tight reservoirs with a relatively high matrix permeability (0.5 mD), the depletion period can be shortened, and the measures for supplementing the reservoir energy can be adopted as soon as possible to further improve single-well EUR. For the reservoirs with extremely low matrix permeability, multiple huff-n-puff cycles are conducive to improve single-well EUR.

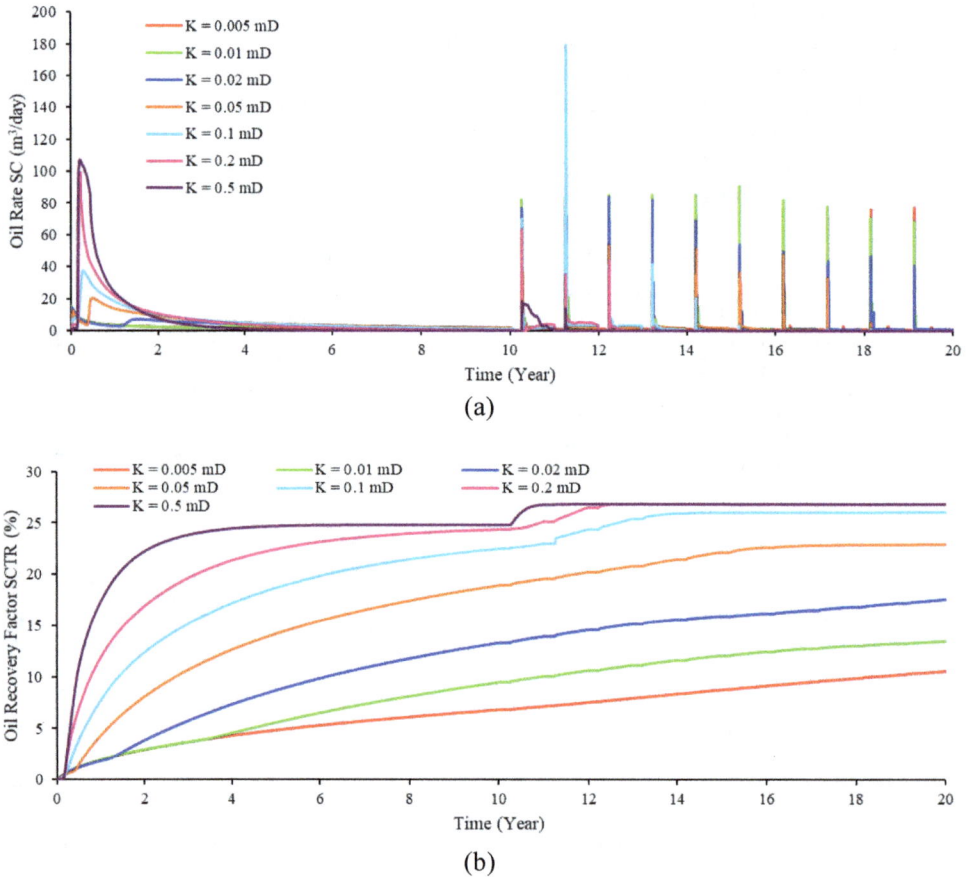

Figure 6:   Effect of matrix permeability on the performance of CO_2 huff-n-puff. (a) Oil rate; and (b) Oil recovery factor.

3.4 Lower limits of coupling physical properties in the economic limit output

Based on the sensitivity analysis of matrix permeability on the performance of CO_2 huff-n-puff, it appears that there are differences in the applicability and performance of CO_2 huff-n-puff under different geological conditions. Therefore, the lower limits of coupling matrix permeability and initial oil saturation for the applicability of CO_2 huff-n-puff are determined, as shown in Fig. 7. Considering the operation costs such as drilling and fracturing, the lower limits of matrix permeability and initial oil saturation are 0.04 mD and 0.52, respectively, if the single-well EUR need reach 20% to achieve profitability. Fig. 8 is the lower limit during the CO_2 huff-n-puff period under different reservoir conditions. When the matrix permeability is in the range of 0.01 mD to 0.1 mD, the performance on improving single-well EUR of CO_2 huff-n-puff is better, and the oil recovery factor increases as the initial oil saturation increases. When the matrix permeability is greater than 0.1 mD, on the one hand, the residual oil saturation after depletion is low; on the other hand, the gas channelling is considerable resulting in poor performance of CO_2 huff-n-puff. As the matrix permeability is less than 0.01 mD, due to the limited injectability and swept area of CO_2, the re-pressurization plays a primary role in improving the single-well EUR and the performance is poor. Infill drilling might be a potential method by further stimulating the reservoirs near the producer.

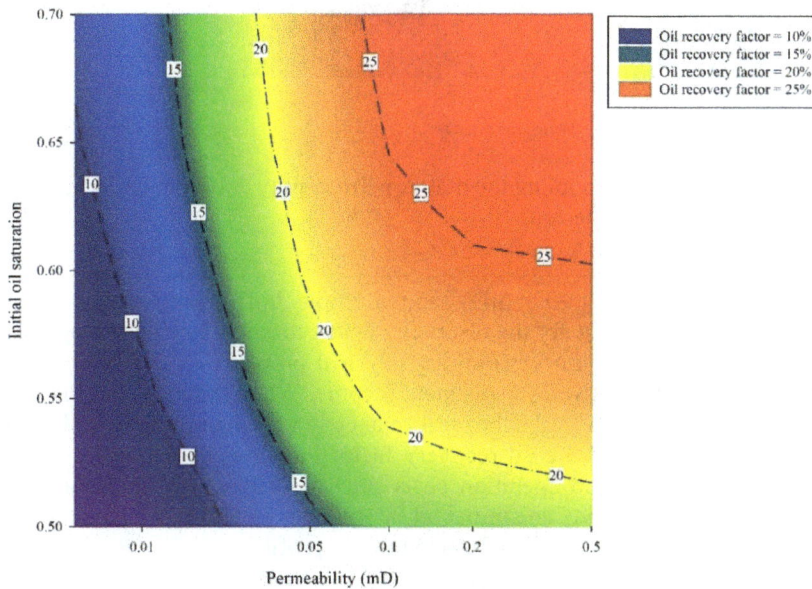

Figure 7:   Lower limits of coupling the matrix permeability and initial oil saturation during the whole development.

4 CONCLUSIONS

In this work, numerical simulations are conducted to investigate the influence of horizontal well length, huff-n-puff cycle, and matrix permeability in improving single-well EUR of CO_2 huff-n-puff. Lower limits of coupling the matrix permeability and initial oil saturation are also determined in the economic limit output. The conclusions are as follows:

1. Longer horizontal well shows a much better performance in improving single-well EUR both in the depletion and CO_2 huff-n-puff stage.
2. There is an optimum injection time for CO_2 huff-n-puff under different reservoir conditions. The depletion period can be shortened in tight reservoirs with a relatively

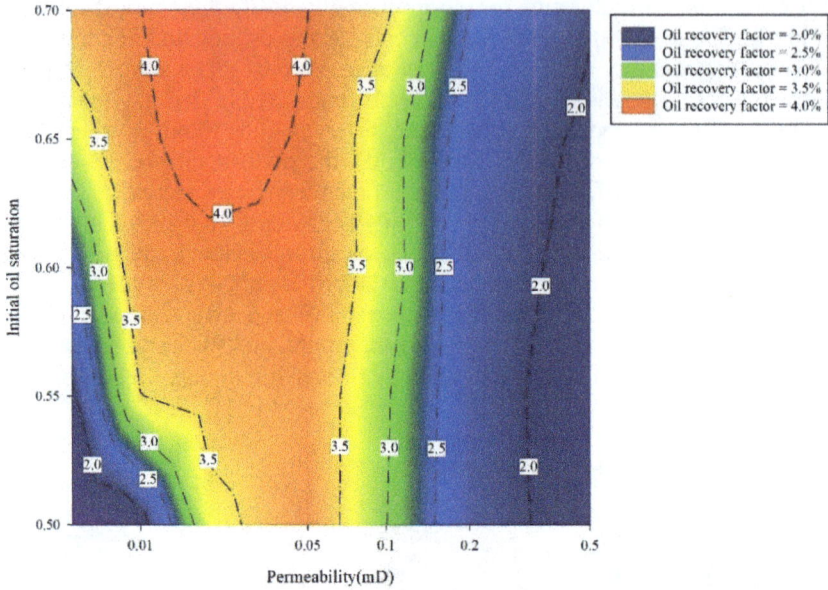

Figure 8:   Lower limits of coupling the matrix permeability and initial oil saturation during CO_2 huff-n-puff period.

high matrix permeability (0.5 mD), and multiple huff-n-puff cycles conducive to improved single-well EUR for the reservoirs with extremely low matrix permeability.

3. Determining the lower limit of coupling physical properties in shale oil reservoirs is critical from the perspective of investment income. When the matrix permeability is in the range of 0.01 mD to 0.1 mD, the performance in improving single-well EUR of CO_2 huff-n-puff is better, and the oil recovery factor increases as the initial oil saturation increases. Considering the operation costs such as drilling and fracturing, the lower limits of matrix permeability and initial oil saturation are 0.04 mD and 0.52, respectively, if the single-well EUR need reach 20% to make a profit.

ACKNOWLEDGEMENTS

The authors acknowledge that this work was partially funded by the Strategic Cooperation Technology Projects of CNPC and CUPB (No. ZLZX2020-02-04-04). We also thank the National Natural Science Foundation of China (No. U52174039) for financial support.

REFERENCES

[1]   Milad, M. et al., Huff-n-puff technology for enhanced oil recovery in shale/tight oil reservoirs: Progress, gaps, and perspectives. *Energy and Fuels*, **35**, pp. 17279–17333, 2021.

[2]   Li, L., Su, Y. & Sheng, J.J., Investigation of gas penetration depth during gas huff-n-puff EOR process in unconventional oil reservoirs. *SPE Canada Unconventional Resources Conference*, p. 26, 2018.

[3]   Jia, B., Tsau, J.S. & Barati, R., A review of the current progress of CO_2 injection EOR and carbon storage in shale oil reservoirs. *Fuel*, **236**, pp. 404–427, 2019.

[4]   Liu, J. & Chapman, W.G., Thermodynamic modeling of the equilibrium partitioning of hydrocarbons in nanoporous kerogen particles. *Energy and Fuels*, **33**, pp. 891–904, 2019.

[5]   Chen, H. et al., A dynamic model of CO_2 diffusion coefficient in shale based on the whole process fitting. *Chemical Engineering Journal*, **428**, 2022.

[6]   Alfarge, D., Wei, M. & Bai, B., Analysis of IOR pilots in Bakken formation by using numerical simulation. *Abu Dhabi International Petroleum Exhibition and Conference*, 2017.

[7]   Clark, A.J., Determination of recovery factor in the Bakken formation, Mountrail County, ND. *SPE Annual Technical Conference and Exhibition*, New Orleans, Louisiana, 2009.

[8]   Alfarge, D., Wei, M. & Bai, B., A parametric study on the applicability of miscible gases based EOR techniques in unconventional liquids rich reservoirs. *SPE Canada Unconventional Resources Conference*, 2018.

[9]   Burrows, L.C. et al., A literature review of CO_2, natural gas, and water-based fluids for enhanced oil recovery in unconventional reservoirs. *Energy and Fuels*, **34**(5), pp. 5331–5380, 2020.

[10] Wan, T. & Sheng, J.J., Enhanced recovery of crude oil from shale formations by gas injection in zipper-fractured horizontal wells. *Petroleum Science and Technology*, **33**(17–18), pp. 1605–1610, 2015.

[11] Pankaj, P. et al., Boosting oil recovery in naturally fractured shale using CO_2 huff-n-puff. *SPE Argentina Exploration and Production of Unconventional Resources Symposium*, p. 15, 2018.

[12] Yan, J., Carbon capture and storage (CCS). *Applied Energy*, **148**, pp. A1–A6, 2015.

[13] Pu, W. et al., Experimental investigation of CO_2 huff-n-puff process for enhancing oil recovery in tight reservoirs. *Chemical Engineering Research and Design*, **111**, pp. 269–276, 2016.

[14] Wan, T. et al., Compositional modeling of EOR process in stimulated shale oil reservoirs by cyclic gas injection. *SPE Improved Oil Recovery Symposium*, 2014.

[15] Lei, Z. et al., Simulation and optimization of CO_2 huff-n-puff processes in tight oil reservoir: A case study of Chang-7 tight oil reservoirs in Ordos basin. *SPE Asia Pacific Oil and Gas Conference and Exhibition*, 2018.

[16] Hoffman, B.T. & Rutledge, J.M., Mechanisms for huff-n-puff cyclic gas injection into unconventional reservoirs. *SPE Oklahoma City Oil and Gas Symposium*, p. 13, 2019.

[17] Zhou, X. et al., Feasibility study of CO_2 huff 'n' puff process to enhance heavy oil recovery via long core experiments. *Applied Energy*, **236**, pp. 526–539, 2019.

[18] Kong, B., Wang, S. & Chen, S., Simulation and optimization of CO_2 huff-and puff processes in tight oil reservoirs. *SPE Improved Oil Recovery Conference*, 2016.

[19] Computer Modeling Group, GEM Manual, 2019.

[20] Wang, P. et al., Modeling phase behavior of nano-confined fluids in shale reservoirs with a modified Soave-Redlich-Kwong equation of state. *Chemical Engineering Journal*, **433**(3), 2022.

EVALUATION OF THE STATE OF THE REGION'S ENERGY COGENERATION SYSTEMS BASED ON THE RELIABILITY OF POWER SUPPLY IN CONDITIONS OF HIGH ECONOMIC RISKS

ALEXEY Y. DOMNIKOV, MICHAEL KHODOROVSKY & LIUDMILA DOMNIKOVA
Academic Department of Banking and Investment Management,
Ural Federal University named after the First President of Russia B.N. Yeltsin, Russia

ABSTRACT

The reliable operation of power cogeneration systems as the most important component of large power systems is essential for national economy's successful development. The timeliness of studying the effective administrative actions to increase the economic effect for energy companies and maintain a high level of readiness of energy systems for overcoming the threats to their sustainable operation, that appear in conditions of high economic risks, is preconditioned by the presence of not only technical but also economic aspects of energy supply reliability. The structural complexity of the regional power cogeneration systems as well as availability of various interrelations between power stations and their performance characteristics lead to the development of accidents and their turning into system ones. This circumstance puts among the most urgent the requirement of a specific study of the problem of reliability analysis of the region's territories' energy supply. In the course of research, the structural systematic characteristics of power cogeneration systems and conditions contributing to appearance of system accidents leading to the loss of sustainability were taken into account. A method of indicative analysis was applied for studying the issues related to reliability and survivability of energy supply systems. Its specificity was an original methodological approach to calculation of indicators threshold values characterizing the state of energy cogeneration systems in terms of energy supply reliability based on discriminatory analysis. It allowed conducting a series of simulation calculations by indicative blocks of a situation belonging to a certain class of states and obtaining generalized assessments characterizing the reliability level of energy cogeneration systems on the territories of the regions.

Keywords: power industry, efficiency, strategy, reliability, ecology, mathematical economic models, uncertainty.

1 INTRODUCTION

Recently observed serious growth of economic risks leads to changing the operating conditions and energy cogeneration systems' management principles. This circumstance causes a substantial decrease of energy supply's reliability levels and actualizes the tasks of studying the issues of management decision making for crisis developments neutralization. It is worth mentioning that the related problems are also those of technical reliability in power cogeneration systems, where reliability means the property of performing given functions in certain amounts under certain operating conditions. At that, reliability is defined as the resultant aggregate of technical characteristics of the facility – the power plants that are a part of distributed and centralized cogeneration systems [1], [2]. Thus, the reliability of power cogeneration systems is related to the ability to avoid attaining a certain limit state (the principle of acceptable damage), after which the system may collapse and be unable to perform its functions even to the baseline minimum necessary for consumers (unacceptable operating parameters, low supply level of energy resources). The property of reliability is intrinsic to power cogeneration systems and manifests itself in abnormal situations. In the context of high economic risks, abnormality in functioning can be triggered not only by factors relating to the technical domain (equipment failure), but also by factors of economic

WIT Transactions on Ecology and the Environment, Vol 255, © 2022 WIT Press
www.witpress.com, ISSN 1743-3541 (on-line)
doi:10.2495/EPM220021

(lack of investment resources, financial sustainability), environmental (environmental pollution, harmful environmental impacts) and socio-political (national and regional conflicts, strikes, etc.) nature. That is why when considering the issues of power engineering operation and development, it is not anymore sufficient to confine oneself to technical aspects of reliability analysis, but one should do a complex research of cogeneration systems reliability – reliability of operation including all above mentioned factors [3].

Although individual events related to the failure of cogeneration systems reliability appear to be random, the impact of negative factors as a whole is not random, but is the natural result of their accumulation, i.e. it has an integral character. So, when assessing the levels of energy supply reliability, first of all, along with the structural ones, it is necessary to use the integral factors of their activity, which allows obtaining more objective results of the research.

2 THE METHODOLOGICAL APPROACH TO CARRYING OUT CLASSIFICATION OF THE STATES OF POWER COGENERATION SYSTEMS BY RELIABILITY LEVELS OF ENERGY SUPPLY WITH THE HELP OF DISCRIMINATORY ANALYSIS

The classification of the states of power cogeneration systems by reliability levels of energy supply was carried out using the analysis system of multivariable data on the basis of discriminatory analysis applied in the theory of image identification. As is known, discriminant analysis is a branch of computational mathematics representing a set of statistical analysis methods for solving the tasks of pattern recognition used to decide which variables separate (i.e. "discriminate") the emergent data sets combined into groups [4], [5].

The general idea of the developed methodological approach is as follows. If the indicator values for the different states of regional energy supply's reliability are known, a training sample can be generated based on the statistical data, containing objects of different state classes of the region: Normal (*N*), Medium (*M*), Low (*L*). The classification of the current state of the regions may be carried out by the indicator values, which characterize energy supply reliability through certain decision rules – discriminator functions.

Using the principle of dichotomy, the objects in the training sample belonging to a given energy supply reliability class are separated from those in the other classes. The character of the discriminator function $E(X)$ carries information about the situation class, and its value carries information about the closeness of the situation to the boundary separating objects of different classes, that is a kind of danger degree of the situation in terms of reliability [5], [6].

If $E(x_1, x_2, \dots, x_m) = 0$, then the point is on the dividing surface. A probabilistic approach, widely used in image identification theory, is used to determine the decision rules. It corresponds to a case when all the images overlap.

If images X_h of all classes $A(X_h)$ are known, then identification of the objects of unknown affiliation $a(X_0)$ must be done according to the rule:

$$X_0 \in X_h \Rightarrow a(X_0) \in A(X_h), h = \overline{1, m}. \qquad (1)$$

Thus, the task is to build models of the K_h classes from the training sample data, on the basis of which a new object $a(X_0)$ can be identified:

$$X_0 \in K_h \Rightarrow X_0 \in X_h \Rightarrow a(X_0) \in A(X_h), h = \overline{1, m}. \qquad (2)$$

Two types of mistakes are possible at classification: missing a target – P_1 and false alarm – P_2.

The decision rule has to enable minimizing the mathematical expectancy of losses associated with misclassification, that is:

$$F(K) = c_1 q_1 P_1(K) + c_2 q_2 P_2(K) \rightarrow min, \tag{3}$$

where q_1 and q_2 = a priori probabilities of occurrence of objects from the first and second classes; c_1 and c_2 = error rates for assigning objects to two classes.

As is known, the method minimizing the average loss $F(K)$ at given values q and c is called the Bayesian method [1], [7]. According to it, for populations of objects obeying the normal law of distribution, an object with parameters X should be classified as population number one if:

$$\ln(c_1 q_1) - 0.5 \left((X - M_1)^T S_1^{-1} (X - M_1) - \ln|S_1| \right) -$$
$$- \ln(c_2 q_2) - 0.5 \left((X - M_2)^T S_2^{-1} (X - M_2) - \ln|S_2| \right) > 0, \tag{4}$$

where X = vector of variables in indicator space; M_1, M_2 = mathematical expectations of classes 1 and 2; S_1, S_2 = covariance matrices of classes 1 and 2; q_1, q_2 = a priori probabilities of occurrence of objects from the first and second classes; c_1 c_2 = prices of incorrect assignment of objects to classes 1 and 2.

Functionals G_i defined by equation $G_i = \ln(c_1 q_1) - 0.5 \left((X - M_1)^T S_1^{-1} (X - M_1) - |S_1| \right)$ are called quadratic informants. In terms of informants, the classification rule looks as follows: the object should be assigned to the population for which its informant is bigger.

In the course of the research, it was found out that the most important disadvantage of using discriminant analysis method for classification of states according to regional energy supply's reliability levels is the difficulty in obtaining samples of statistically significant size.

As mentioned above, the classification of regions and their energy cogeneration systems is carried out with the help of thresholds for indicators characterizing the reliability of power supply. From this point of view, a situation existing in a certain region at a given moment may be classified for example as Normal (N), Medium (M), Low (L). The thresholds separating one class from another by some indicator may be marked as X_M and X_L. X_M helps to separate class N from M, and X_L, correspondingly – M from L.

According to the developed method, separation surfaces of the considered pairs of classes are built in the space of indicators. To construct them, a training sample consisting of points with known states of energy supply reliability is used [3]. Threshold points are then defined on these surfaces as intersections with the lines connecting the class centroids.

The equation of the surface separating classes 1 and 2 is written as follows:

$$\ln(c_1 q_1) - 0.5 \left((X - M_1)^T S_1^{-1} (X - M_1) - \ln|S_1| \right) -$$
$$- \ln(c_2 q_2) - 0.5 \left((X - M_2)^T S_2^{-1} (X - M_2) - \ln|S_2| \right) = 0, \tag{5}$$

A line passing through the first and second class centroids with coordinates M_1 and M_2 is written as follows:

$$X = b(M_2 - M_1) + M_1, \tag{6}$$

where b is the straight line parameter.

By substituting (6) into (5), we obtain an equation with respect to the parameter b:

$$\ln(c_1 q_1) - 0.5 \left((b(M_2 - M_1) + M_1 - M_1)^T S_1^{-1} (b(M_2 - M_1) + M_1 - -M_1) - \ln|S_1| \right)$$
$$- \ln(c_2 q_2) - 0.5 \left((b(M_2 - M_1) + M_1 - -M_2)^T S_2^{-1} (b(M_2 - M_1) + M_1 \right.$$
$$- M_2) - \ln|S_2| \right) = 0. \tag{7}$$

After algebraic transformations, this equation is reduced to a standard form of quadratic equation:

$$b^2 A_1 + b A_2 + A_3 = 0, \tag{8}$$

where

$$A_1 = 0.5(M_2 - M_1)^T (S_1^{-1} + S_2^{-1})(M_2 - M_1), \tag{9}$$

$$A_2 = (M_2 - M_1)^T S_2^{-1}(M_2 - M_1), \tag{10}$$

$$A_3 = -0.5(M_2 - M_1)^T S_2^{-1}(M_2 - M_1) - \ln \frac{c_1 q_1}{c_2 q_2} + 0.5 \ln \frac{|S_2|}{|S_1|}. \tag{11}$$

One of two roots satisfying condition $0 \le b_0 \le 1$ corresponds to the intersection point of the straight line and the separating surface on the segment between the classes centroids. Using it and relation (6), the threshold values for safety indicators between classes 1 and 2 are determined.

$$X_0 = b_0(M_2 - M_1) + M_1. \tag{12}$$

3 EVALUATION OF THE STATE OF POWER ENGINEERING SYSTEMS IN THE URALS REGION IN TERMS OF ENERGY SUPPLY RELIABILITY

The Urals region is located on the Middle, the South, and partially on the North Ural, as well as on adjacent parts of the East-European and the West-Siberian plains at an area of 823.3 thousand km². It includes seven subjects of the Russian Federation: the Sverdlovsk Oblast, the Perm Krai, the Chelyabinsk Oblast, the Kurgan Oblast, the Orenburg Oblast, the Udmurt Republic and the Republic of Bashkortostan. The Urals region is exceptionally rich in various minerals. It is a highly developed heavy industry production area with a complex structure. Mineral raw materials and gas extraction, logging and timber processing are of nationwide importance. The district's industry is particularly characterized by a high level of production concentration, intra- and inter-sectoral cooperation and combination, as well as by a well-developed infrastructure, including electric power engineering. Ferrous metallurgy is the primary industry of the Urals region. Mechanical engineering (energy, transport, agriculture), forestry, chemicals, petrochemicals and mining, as well as oil and gas extraction and processing are sufficiently well developed industries.

In the framework of their research, the authors successfully apply the indicative approach for assessment of reliability levels. To assess reliability levels, this approach involves the use of some combinations of indicators – immediately assessed initial parameters [8]–[10]. Further on, these indicators are grouped in units that reflect some characteristic features in operation of power cogeneration systems. All indicators in the handled problem are divided in six units. The list of units, indicators, and threshold values is given in Table 1.

The main results of solving the set task are given in Table 2. Their analysis shows that the overall reliability of electric power supply on all territories of the Urals region is evaluated as low. It belongs to the class of insufficient reliability to a high degree only in the Sverdlovsk Oblast.

Assessments under the unit of power cogeneration systems' adequacy (unit 1) are within the range of "insufficient reliability and acceptable to some degree" (the Perm Krai) to "definitely low reliability" (the Udmurt Republic). The determining indicator under this unit is the ratio of the sum of the available power plants capacity and the throughput capacity of power links to the maximum power load of consumers (indicator 1.3).

Table 1: The indicators composition and threshold values for assessment of energy supply reliability of the Urals regions.

Name of blocks and indicators	Symbol	Threshold values		
		N	M	L
1. Unit of power cogeneration systems adequacy				
1.1. Share of own sources of electric energy in the energy balance of the territory.	1.1	75	65	55
1.2. Ratio of available power plants capacity to the maximum consumers power load (%)	1.2	100	85%	70
1.3. Ratio of the sum of the available power plants capacity and the throughput capacity of power links to the maximum power load of consumers (%)	1.3	220	190	160
2. Unit of fuel supply systems adequacy				
2.1. Share of own sources in the balance of boiler and furnace fuels (%)	2.1	70	50	30
2.2. Share of own sources in the balance of motor fuel (%)	2.2	75	55	35
2.3. Share of completion of the target on coal accumulation,%	2.3	100	90	80
2.4. Share of completion of the target on black oil fuel accumulation (%)	2.4	90	75	60
3. Unit of power cogeneration systems structural and performance reliability				
3.1. Ratio of the reserved capacity of power plants to consumers' maximum electrical load (%)	3.1	14	10	6
3.2. Availability factor of the electric power plants generating equipment.	3.2	86	78	70
4. Unit of power cogeneration systems survivability				
4.1. Share of predominant heat resource in the consumption of boiler and furnace fuels (%)	4.1	35	50	65
4.2. Installed capacity share of the largest power plant (%)	4.2	25	35	45
4.3. The level of securing the demand for heat sources capacity in context of a sharply increased demand	4.3	105	95	85
5. Unit of power companies main production assets reliability and efficiency				
5.1. Wear-out rate of electric power companies main production assets (%)	5.1	30	45	60
5.2. Depreciation degree of the fuel industry enterprises basic production assets,%	5.2	30	45	60
5.3. Share of commissioning the installed capacity and technical re-equipment of electric power plants in the territory over a five-year period (%)	5.3	8	5	2
6. Unit of assessment of financial and economic performance of electric power companies				
6.1. Ratio of the excess of overdue accounts payable of the electric power industry enterprises of the territory over the overdue accounts receivable to their annual production volume (%)	6.1	5	10	15
6.2. Ratio of the excess of overdue accounts payable of enterprises in the fuel industry of the territory over the overdue accounts receivable to their annual output (%)	6.2	4	8	12
6.3. Ratio of overdue accounts payable of electric power enterprises to their annual output (%)	6.3	20	35	50
6.4. Ratio of overdue accounts payable of fuel industry enterprises to their annual output (%)	6.4	20	35	50

Table 2: The results of assessment of the status of power cogeneration systems of the Urals region by power supply reliability.

Region name	RC	Indicator block number												Overall rating	
		1		2		3		4		5		6			
		DS	DI	DS	DI	DS	DI	DS	DI	DS	DI	DS	DI	DS	DI
Perm Krai	N	0.865	1.3	0.958	2.1	1.000	3.2	0.002	4.1	0.002	5.1	0.109	6.2	0.002	4.1,
	M	1.000	1.3	0.989	2.1	0.841	3.1	0.230	4.1	0.995	5.1	0.193	6.2	0.193	6.2
	L	2.293	1.3	1.000	2.1	0.445	3.1	1.000	4.1	1.000	5.3	1.000	6.2	1.000	2.1, 4.1, 5.3, 6.2
Sverdlovsk Oblast	N	0.088	1.3	0.989	2.1	0.473	3.2	0.502	4.1	0.079	5.1	0.971	6.3	0.079	5.1
	M	0.995	1.3	1.000	2.1	0.971	3.2	1.000	4.1	1.000	5.1	1.000	6.2	0.971	3.2
	L	1.000	1.3	0.999	2.1	1.000	3.2	0.791	4.1	0.908	5.3	0.999	6.2	1.000	1.3, 3.2
Orenburg Oblast	N	0.003	1.3	1.000	2.1	0.531	3.2	0.003	4.1	0.002	5.1	0.763	6.3	0.001	1.3, 4.1
	M	0.035	1.3	0.958	2.1	0.989	3.2	0.109	4.1	1.000	5.1	0.907	6.2	0.035	1.3
	L	1.000	1.3	0.908	2.1	1.000	3.2	1.000	4.1	0.981	5.1	1.000	6.1	1.000	1.3, 3.2, 4.1, 6.1
Republic of Bashkortostan	N	0.365	1.3	1.000	2.1	0.908	3.2	0.003	4.1	0.005	5.2	0.908	6.3	0.003	4.1
	M	1.000	1.3	0.999	2.1	1.000	3.2	0.316	4.1	0.817	5.2	1.000	6.3	0.316	4.1
	L	0.791	1.3	0.981	2.1	0.865	3.2	1.000	4.1	1.000	5.3	0.865	6.3	1.000	4.1, 5.3
Udmurt Republic	N	0.088	1.3	0.943	2.2	0.943	3.2	0.009	4.1	0.098	5.1	0.989	6.3	0.009	4.1
	M	0.995	1.3	0.999	2.2	1.000	3.2	0.473	4.1	0.791	5.3	1.000	6.3	0.473	4.1
	L	1.000	1.3	1.000	2.2	0.817	3.2	1.000	4.1	1.000	5.3	0.943	6.3	1.000	1.3, 2.2, 4.1, 5.3
Chelyabinsk Oblast	N	0.002	1.3	1.000	2.4	0.316	3.2	0.001	4.1	0.071	5.2	0.002	6.2	0.002	6.2
	M	0.044	1.3	0.560	2.4	0.865	3.2	0.177	4.1	0.865	5.3	0.004	6.2	0.004	6.2
	L	1.000	1.3	0.316	2.4	1.000	3.2	1.000	4.1	1.000	5.3	1.000	6.2	1.000	1.3, 3.2, 4.1, 5.3, 6.2
Kurgan Oblast	N	0.009	1.3	0.971	2.4	1.000	3.2	0.473	4.1	0.391	5.2	0.003	6.3	0.009	6.3
	M	0.619	1.3	1.000	2.4	0.989	3.2	1.000	4.1	1.000	5.2	0.005	6.3	0.005	6.3
	L	1.000	1.3	0.995	2.4	0.531	3.2	0.817	4.1	0.589	5.3	1.000	6.3	1.000	1.3, 6.3

Note: RC = Reliability characteristic; DS = the degree of situation belonging to a certain class of states; DI = determining indicator, i.e. indicator under which the situation is related to a certain class.

Under the second unit, three regions (the Perm Krai, the Udmurt Republic and the Kurgan Oblast) get low or insufficient reliability assessments. Other areas are recognized more or less safe. An opposite situation is observed under the unit of power generation systems' structural and performance reliability. (unit 3). The mentioned areas have high or acceptable assessments and the rest of them – insufficient or low ones, and it is determined by the availability factor of the electric power generating equipment.

The unit of power cogeneration systems survivability (unit 4) is characterized by low assessments for all areas due to the strongly pronounced fuel balance mono-structure (indicator 4.1). Additionally, one can also note the poor assessments of the level of securing the demand for heat sources capacity in context of a sharply increased demand (indicator 4.3) for the Perm Krai and the Sverdlovsk Oblast.

The worst situation from the point of view of power supply's reliability has formed in the unit of power cogeneration systems' efficiency (unit 5). All indicators of this unit for all areas are at low or insufficient level. Along with a high wear degree of the main production facilities, rather low rates of new facilities putting in operation and power objects modernization are observed. This situation can be explained by a low investment attractiveness of power engineering, mainly due to long payback periods.

Another reason of the relatively poor state of main production facilities becomes visible at the analysis of the results of unit 6, reflecting the financial and economic performance of power companies. According to this unit, all the regions of the Urals get unsatisfactory assessments, and this is connected with the high level of accounts payable of power companies.

4 CONCLUSION

The complexity of cogeneration systems in the Urals region, the difficulties in their management, the complex interrelationships between energy facilities and their regime indicators may lead to conditions for the development of accidents and their turning into system accidents. This is confirmed by the requirement of a special consideration for the task of analyzing the reliability of energy supply of the regions based on tracking the structural systematic characteristics and conditions that contribute to appearance of system accidents leading to the loss of survivability by energy cogeneration systems.

The method of estimating indicator thresholds based on discriminant analysis has been developed and tested for the purpose of indicative analysis. It allowed obtaining the threshold indicator values from training samples and making a comprehensive assessment of the state of energy cogeneration systems in the Ural region in terms of energy supply reliability.

In our further research based on the suggested approach, we plan to study the effect of renewable energy sources on the system reliability, and to identify the optimal share of various types of renewable sources capacity that has an impact on the sustainable operation of regional electric power industry.

ACKNOWLEDGEMENT

The work was supported by Russian Foundation for Basic Research (RFBR), contract № 20-010-00886.

REFERENCES

[1] Domnikov, A., Khodorovsky, M. & Domnikova, L., Methodological approach to the research of energy cogeneration systems operational reliability indicators. *International Journal of Energy Production and Management*, **6**(3), pp. 263–276, 2021. DOI: 10.2495/EQ-V6-N3-263-276.

[2]   Domnikov, A., Khodorovsky, M. & Domnikova, L., Identification and classification of the states of cogeneration systems by competitiveness levels of power generating companies. *WIT Transactions on Ecology and the Environment*, vol. 254, WIT Press: Southampton and Boston, pp. 25–32, 2021. DOI: 10.2495/ESUS210031.

[3]   Domnikov, A., Khodorovsky, M. & Domnikova, L., Decision support system used to improve the competitiveness of a power generating company under conditions of uncertainty. *WIT Transactions on Ecology and the Environment*, vol. 254, WIT Press: Southampton and Boston, pp. 15–23, 2021. DOI: 10.2495/ESUS210021.

[4]   Bai, J. & Perron, P., Computation and analysis of multiple structural change models. *Journal of Applied Econometric*s, **18**, pp. 1–22, 2003.

[5]   Lachenbruch, P., *Discriminant Analysis*, Hafner Press: New York, 458 pp. 1995.

[6]   Stevens, J., *Applied Multivariate Statistics for the Social Sciences*, Lawrence Erlbaum: New Jersey, 734 pp. 1996.

[7]   Jardine, N. & Sibson, R., *Mathematical Taxonomy*, John Wiley: London, 1971.

[8]   Mehigan, L., A review of the role of distributed generation (DG) in future electricity systems. *Energy*, **163**, pp. 822–836, 2018. DOI: 10.1016/j.energy.2018.08.022.

[9]   Johnstone, P. & Kivimaa, P., Multiple dimensions of disruption, energy transitions and industrial policy. *Energy Research and Social Science*, **37**, pp. 260–265, 2018. DOI: 10.1016/j.erss.2017.10.027.

[10]   Scholten, D. & Künneke, R., Towards the comprehensive design of energy infrastructures. *Sustainability*, **8**(12), pp. 1291–1295, 2016. DOI: 10.3390/su8121291.

ADOPTION OF TECHNOLOGY PLATFORMS IN THE ELECTRIC POWER INDUSTRY: NEW OPPORTUNITIES

LAZAR D. GITELMAN & MIKHAIL V. KOZHEVNIKOV
Academic Department of Energy and Industrial Enterprises Management Systems,
Ural Federal University, Russia

ABSTRACT

The paper defines the key areas of application and economic effects of platform tools in the electric power industry. This is a new approach that boosts the development of knowledge-intensive services on the basis of integration of utilities, consumers, developers and digital solutions providers into a single eco-system, which enables the accumulation and update of knowledge on the systems and processes being serviced. Using theoretical analysis and their specific hands-on experience, the authors prove that the application of platform tools delivers an economic effect through a greater volume and speed of energy transactions, the adoption of innovations by energy companies in their commercial and production operations, faster transfer of new knowledge between platform stakeholders, and creation of knowledge-intensive services for pro-active consumers on the basis of new knowledge. The scientific novelty of the study is that it gives a systemic overview of theoretical and methodological conceptions of platforms tailored for the electric power industry, which is a complex infrastructure industry, from the perspective of the economic expediency of their implementation. The findings could be of interest to various parties involved in energy and economic relations when implementing projects focusing on modernization, energy transition, energy conservation, improvement of consumer interactions with the employment of the latest communication interfaces.

Keywords: electric power industry, technology platform, knowledge-intensive service, business model, demand side management, transaction.

1 INTRODUCTION

The widespread adoption of the technology platform concept is closely associated with the accelerated development of innovations in many countries and stronger competition in research intensive markets. The new reality implies that the creation and adoption of breakthrough innovations in high-tech industries is impossible without the close cooperation of various "agents" – companies, universities, research and science centers – and without the development of effective mechanisms for the transfer of knowledge and technology or the creation of strategic alliances, including at the industrial and state levels.

In Gitelman et al. [1], the authors prove that roughly all interpretations of the concept "technology platform" match one of two approaches. Under the first one, a technology platform is a self-regulating consortium that is created for solving super complex problems, typically of scientific and technical nature. Problems that are undertaken by such platforms are always unique and one-of-a-kind, new and non-linear, while lacking a sufficient bank of knowledge, especially in cross-disciplinary fields, and the essential analytical base.

More popular, though, is the so-called "product" or "market" approach that suggests looking at the technology platform as a starting ground for businesses seeking to build their own eco-system (a new market segment). According to this approach, a large company acts as a future technology provider, with numerous smaller contractors, research centers, various technological and logistics services and, most importantly, consumers gravitating toward it. The key "product" of the platform are services (or, more rarely, tangible products) that are built upon unique technological architecture [2].

WIT Transactions on Ecology and the Environment, Vol 255, © 2022 WIT Press
www.witpress.com, ISSN 1743-3541 (on-line)
doi:10.2495/EPM220031

Within the framework of the second approach, a technology platform can be defined as a peculiar system for organizing technical, information and intellectual assets in a digital environment. By using the platform, participants are able to jointly create streams of bundled products and services that provide unique consumer value that can be modified in line with customer preferences by quickly adding new components to the products or removing unneeded ones [3].

By using the second approach, many high-tech companies were able to become global leaders by market capitalization. As a result, it is platform companies who boast the highest business value, outpacing significantly traditionally built companies (Fig. 1).

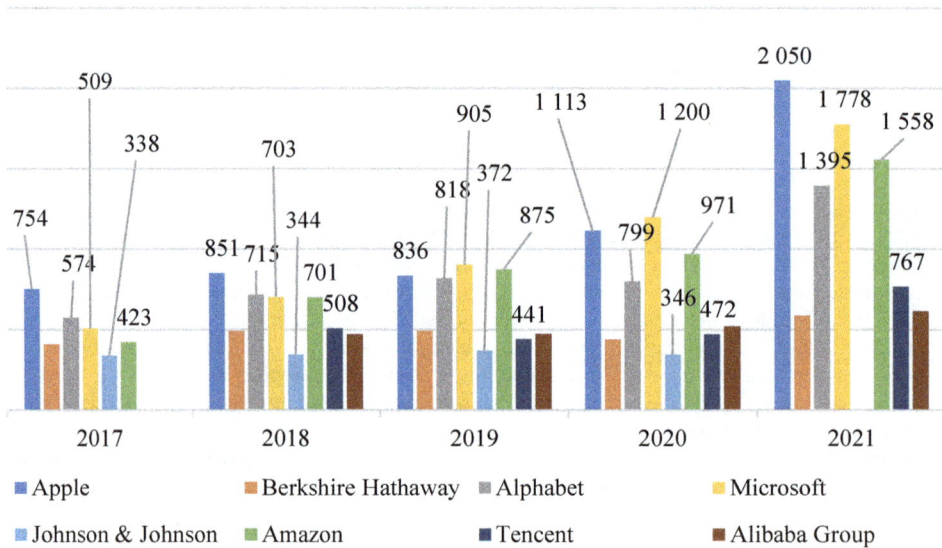

Figure 1: Development dynamics of digital platform-based businesses (billion dollars).

The pioneers of platform business models are companies, such as Apple or Amazon, that essentially do not own any tangible assets. In the past few years, though, the trend towards platform-based business models has emerged in a number of heavy industry sectors (Table 1).

Table 1: Industries with the highest potential for embracing platforms (%) [4].

Industry	Global	USA	China	India	Japan	UK
Automotive/transportation	13	13	18	14	13	16
Consumer markets/retail	11	13	14	9	7	8
Education	8	13	8	5	13	5
Energy	8	5	6	14	0	10
Healthcare	12	11	11	6	17	11
Industrial manufacturing	10	10	4	8	13	15
Media	14	18	11	6	13	18
Telecom	10	7	16	12	23	5

Among the factors that warrant the creation of platforms in the electric power industry there are trends towards decentralization, the expansion of RES, customer centricity, integration between energy and other urban infrastructure sectors, the emergence of prosumers – those who both consumer and produce electricity [5], [6]. Platforms are gradually finding demand in three neighboring industries, yet they differ in terms of the core technologies that form the foundation of their architecture (Table 2). In the first segment, key transformations take place within grid companies and electricity resellers that have a radically higher degree of interaction with end users, the system operator, dispatch structure, and energy service companies (ESCOs) thanks to smart meter technologies. In the energy-as-a-service segment, energy companies – and primarily energy resellers – a transition is under way to a model of selling packages of knowledge-intensive services (e.g., creation of customized microclimate or integrated management of utilities infrastructure) based on energy consumption management by means of adaptive information and communications technologies. Finally, platforms are taking over the electric vehicle (EV) industry, calling for the emergence of new market agents – independent organizations or subsidiaries of energy companies that would manage energy flows between energy consuming devices and the grid. The emergence of platforms is, therefore, a major transformative force for the business models of energy and service businesses.

Table 2: Examples of emerging platforms in the electric power industry.

Segment	Basic technology	Key participants	User groups
Smart grids	Smart metering systems	• Grid operator • Power dispatcher • ESCOs	Large power companies Power consumers
Energy-as-a-service	Adaptive information and telecom systems	• ESCOs • Specialist telecom companies • Data processing centers • Suppliers of auxiliary knowledge-intensive service (finances, commerce, consulting)	Retail electricity providers or independent power producers Electricity consumers
Electric vehicles	Charging network coupled with energy storage and accumulation systems	• Specialist company providing maintenance for charging infrastructure • Demand side response aggregator • Telecommunication company • Data processing center	Retail electricity providers Power dispatcher Power consumers

Some studies [7], [8] say, however, that the electric power industry does not take full advantage of platforms' potential, which is due to a shortage of the necessary methodological tools and a comprehensive review of actual experience, especially from the point of view of the economic expediency of their implementation. This study aims to produce a systemic overview of promising trends and to substantiate potential economic effects of the implementation of platform solutions that use the latest technologies, including from the field of digitalization, in the electric power industry.

2 METHODOLOGICAL FOUNDATIONS FOR ECONOMIC INTERACTION BETWEEN PLATFORM STAKEHOLDERS

The literature on platforms contains two contradicting views as to the arrangement of economic relations among platform stakeholders [9], [10].

2.1 View 1

Platforms must be viewed as participating in several one-sided markets, that is, each side must be considered as interacting with the platform in a separate market. Analysis of effects on competition assumes that any action that makes the situation worse for users in any of the markets, such as a price hike enforced by the platform for one of the sides, constitutes a sufficient ground for considering the platform's actions as undermining competition.

As the platform provides its service to each of the sides separately, it is logical to look at two separate markets in which the platform provides information to participants about transaction options. This, however, reveals a "zero price problem", meaning that one of the groups of participants could connect to the platform free of charge and avoid paying for its services. The connection of one of sides to the platform is de facto subsidized by another side. In that case, many competitive analysis tools do not work because it becomes impossible to analyze price increases. Meanwhile, zero or negative pricing could be the platform's deliberate strategy because, for example, allowing users on one side to connect to the platform free of charge produces positive indirect external effects (by increasing the value of immediate contact through platform) for the subsidizing side. All else being equal, the pricing structure offered by the platform is neutral, while the price set for one of the groups of users connecting to the platform depends on the price being paid by users from another group.

2.2 View 2

Platforms are naturally designed to build connections between sides, so in order to analyze the boundaries of the commodity market, it is necessary to consider all sides connecting to the platform, and indirect network effects. The key parameter for analysis is the presence of the "transfer pricing effect" accompanying transactions: one of the sides connecting to the platform can change the prices and, by doing so, fully pass the changes to the price set by the platform on to the counterparties it interacts with through the platform. If the effect is observed, for the sake of competitive analysis it would be advisable to study separate markets where the platform sets prices for each of the groups of users being studied. This is due to the fact that platforms that face the price transfer effect are unable to internalize emerging indirect effects by changing their pricing structure, which means their multi-sided business model does not play any role in price setting.

One can decide whether to examine one multi-sided market or several interconnected markets where no price transfer effect is observed by looking at whether or not the platforms allow transactions between the sides. In this regard, one should distinguish between non-transaction matching platforms, non-transaction audience-providing platforms, and transaction platforms that are capable of carrying out transactions between sides and receive a fee for both connection and usage.

If platform users on both sides connect to it in order to find a user on the other side, it is reasonable to say that the platform provides a matching service. In that case, competitive analysis should be performed for one multi-sided market.

If a least one of the user groups connects to the platform in order to access certain content rather than to get information about transaction options, we deal with the case of audience markets that should be considered as several interconnected markets.

A summary of the views explained above is given in Table 3.

A conclusion can be made that the adoption of platforms increases the versatility of the forms of economic interaction between customers and vendors on the basis of emerging direct and indirect network economic effects [12], [13].

Table 3: A summary of views on analysis of market boundaries across platforms [9]–[11].

Indicator	Transaction platform		Non-transaction platform	
Price carryover effect	No	Yes	No	Yes
Matching services	Yes	Yes	Yes	No
Indirect external effects	+	+	+	+/−
Preferred analysis perspective	One multi-sided market	Several one-sided markets	One multi-sided market	Several connected markets

Direct effects are mainly associated with the communication capabilities of platforms that enable them to generate a constant growth of users and transactions thanks to the unique value provided by the platform. As a rule, the value takes the form of access to exclusive offers, analytics or technological architecture that is designed for the production of new services. Access to the platform can be fee-based or free for one or all engaged sides. The economic logic of the platform-based business model is described by the following logical sequence: *user growth → growth in demand for services on offer → growth in prices of services on offer → revenue and profit growth.* The key costs for platform owners are (1) the cost of user acquisition and support; (2) capital and operating costs of developing, creating and maintaining the operability and of the future expansion of the information infrastructure of the platform; (3) associated recruitment costs, the cost of legal support and other platform management processes. The second group – so-called infrastructure costs – includes multiple components and is determined by the technological nature of the platform. It can, for example, depend on the costs of data mining, their transformation and subsequent upload into data centers, data protection costs, analytical query response times [14], the amount of processed information, the types of connection in the network, the structural and functional architecture of the platform [13].

An access fee usually covers operating costs, but it slows down user growth. At the same time, the presence of a "registration fee" reduces the number of unscrupulous users (including spam generators and cyber criminals), so ultimately it helps reduce per-user service costs.

If a platform focuses strictly on matching customers and vendors of various services, i.e., if it acts as a digital middleman, the direct economic effects for the platform owner could be a commission over transactions and revenue from content placed by sponsors or advertising clients.

Considering the above, a general direct economic effect E_{plt} of platform deployment can be calculated by the equation:

$$E_{plt} = \sum_{i=1}^{n} Q_i + \sum_{i=1}^{n} Q_i \cdot K_t + Fee \cdot N - C_i - C_c - C_a, \qquad (1)$$

where Q_i is the financial outcome of the i-th transaction a service, K_t is the percentage of commission on transactions across the platform, *Fee* is the platform access fee, N is the number of platform users, C_i are infrastructure costs, C_C are customer acquisition and support costs, C_a are auxiliary costs.

This equation is usable by a platform owner regardless of whether he or she is also the service supplier (transaction beneficiary) or only works to select vendors and service suppliers. The key difference is that in the second case the first summand is equal to zero.

In order to assess the direct economic effect for service producers and service customers who use a platform as a digital communication tool, one should use modified variants (2) and (3) of the above equation:

$$E_{plt}^{producer} = \sum_{i=1}^{n} Q_i + \Delta C_i + \Delta C_a - \sum_{i=1}^{n} Q_i \cdot K_t - Fee, \qquad (2)$$

where ΔC_i and ΔC_a are the change in infrastructure and auxiliary costs incurred by the producer when he or she sells services independent as opposed to platform involvement;

$$E_{plt}^{customer} = \Delta P + \Delta C_i + \Delta C_a - Fee, \qquad (3)$$

where ΔP is the change (reduction) in the price of the service offered by the platform.

It is worth saying that *indirect economic effects* that emerge in the course of platform interaction are essential for a service customer. They appear when the platform helps the customer increase the efficiency of individual processes (e.g., reduce the cost of service production, boost productivity and operability, reduce the risk of breakdown, receive consulting support). For example, platform tools can enable the users to exchange knowledge and information or provide access to other IoT units for real-time control.

It should be emphasized that a direct effect is only generated by transaction platforms, while non-transaction platforms only produce indirect economic effects.

3 PRACTICAL AREAS OF PLATFORM USAGE IN THE ELECTRIC POWER INDUSTRY

We shall consider areas of platform application in the electric power industry as a promising knowledge-intensive service and take a detailed look at how forms of economic interaction between energy and service companies change as a result of the adoption of platform tools.

3.1 Faster and more transparent energy transactions

Within this trend, transaction platforms help various agents engaged in energy and economic relations to communicate almost in real time while considerably improving the quality of data on energy consumption and related costs. For example, Fig. 2 shows how a digital platform that is integrated into a regional system for power sales management, first of all, delivers on the popular concept of customer centricity. Second, it connects not only all traditional energy market players, but also service providers from various segments in one network; third, it creates an analytical base for addressing comprehensively a number of energy companies' present-day problems that are associated with the technological issues of power supply. Such platforms can be accomplished by means of a three-layer scheme. The first layer of the platform is installed on the consumer's side; the second is installed on the energy supplier's side; and the third one is on the market side. All three layers are connected via a cloud service-oriented interface [15].

Such platforms are capable of delivering current and retrospective consumption data and bill settlements by constructing load charts online. They can pay electricity bills and give suggestions as to possible energy savings throughout the day, coordinate the charging of electric vehicles in a city, advise EV owners on the potential of using EVs as power generators in energy-hungry regions; inform consumers about power outages by sending an app notification; deliver updates on the energy system status (failures, overloads, disruptions, change of rates, installation works, validation of electricity meters); and provide guidance on energy efficient behaviors.

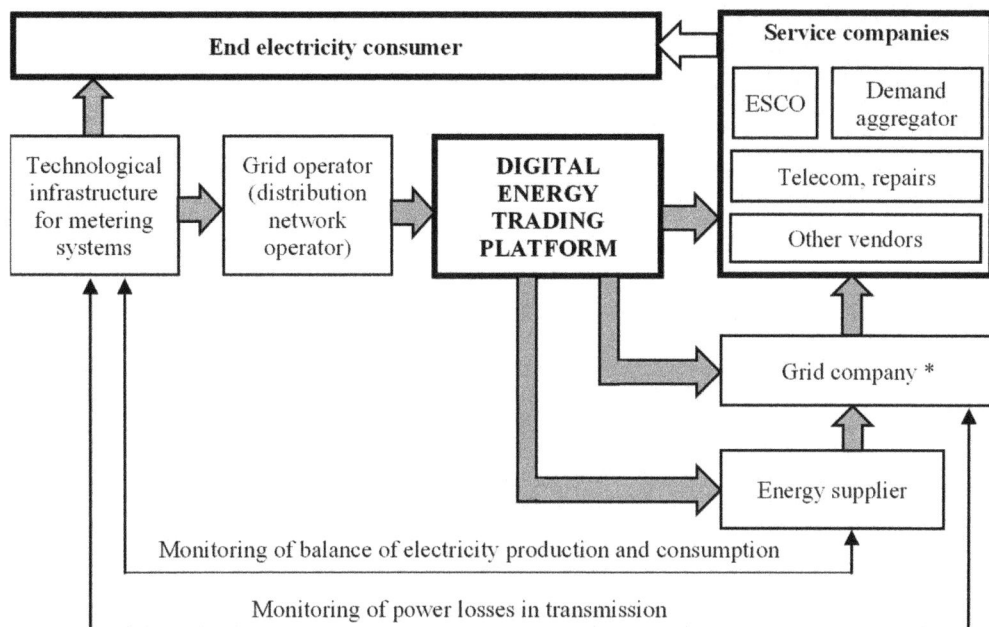

End electricity consumer		Service companies

A model of interaction among energy companies, consumers and service organizations.

Technological infrastructure for metering systems → Grid operator (distribution network operator) → DIGITAL ENERGY TRADING PLATFORM → Service companies (ESCO / Demand aggregator / Telecom, repairs / Other vendors) / Grid company * / Energy supplier

Monitoring of balance of electricity production and consumption

Monitoring of power losses in transmission

* One company can function as both a grid company and an electricity seller

Arrow designations:

«Raw» meter data arrive in real time. No data validation.

Data for billing, calculation of energy consumption balance and service operations. Delays are possible in data transfer due to the time needed for data validation and finalization.

Data for billing and service operations. The most precise data that has been validated and subjected to comparative interpretation to make forecasts and build power consumption models. Data transfer delays are possible.

Figure 2:   A model of interaction among energy companies, consumers and service organizations when platform tools are used for energy transmission.

We shall describe an economic effect that could be achieved by deploying a similar platform that is being designed now by an independent energy seller that operates in a number of Russia's regions. In addition to actual energy supply services, the company installs automated commercial energy metering systems (ACEMS) and conducts energy audits for households as well as industrial customers.

It is supposed that the platform will enable:

1)   an increase in private individuals among the company's subscribers thanks to an application with a customized interface that contains analytical information on current and retrospective power consumption (free of charge). There are plans for enhancing the platform with solutions for "smart home" management and maintenance of smart metering systems and with appropriate energy infrastructure (to be available for a service fee), which would mean a transition to the "energy-as-a-service" business model.

2) an improvement in the quality of energy consumption management and optimization in industrial customers by means of integration between ACEM systems with the platform with an outlook for services that would create a digital twin of the customer's energy supply system.

The company owns the platform and intends to carry out information and technological maintenance of the platform independently. The platform is being developed by a specialist IT company. Immediate (infrastructure) costs C_i are estimated at 12 million roubles (around 160,000 dollars at the current exchange rate); the expenditures are to be made in the first year; annual operation costs (to acquire and support customers and carry out maintenance, $C_c + C_a$) are estimated at 2.4 million roubles in the first year of the project; and 2.6 million roubles in the second year.

Between 2022 and 2023 the company plans to use the platform as a marketing instrument and attract 1,500 new subscribers in the first year and another 3,500 in the second year. This should make it possible to increase electricity sales by about 31.2m kWh and increase revenue by 3.1 million roubles given the estimated retail markup of 0.1 roubles per kWh. Additionally, the company expects to undertake the maintenance of some 100 households with installed "smart home" systems (revenue will be accrued from the platform subscription fee totaling 0.5 million roubles and payments for service works amounting to 5.5 million roubles). Industrial customers are expected to generate 1.2 million roubles in platform access fee and some 10 million roubles for ACEMS installation and maintenance.

The platform rollout costs will, therefore, total 17 million roubles within a period of two years; revenue is projected at 20.3 million roubles and the combined economic effect – at 3.3 million roubles.

In the case of larger energy retailers that act as guaranteeing suppliers and service cities with populations exceeding one million, the economic effect of platform deployment would be much (dozens of times) greater. This is due to the number of subscribers as well as their retail markups that is lower at independent energy retailers.

In the case described above, the transition to a platform-based business model amplifies a number of other direct and indirect economic effects that are achieved thanks to the installation of ACEMSs: lower costs of energy consumption data collection and processing for billing purposes; lower costs of inspection of parameters and quality assurance of energy consumption metering; reduction of business losses and receivables; expanded functionality and greater flexibility of interactions with customers thanks to better understanding of their load schedules and payment structure, which enables the development of more rational end economically reasonable electricity tariffs.

3.2 Demand side management

For the full implementation of this essential mechanism of energy efficiency, regions need to have special technology infrastructure. Digital *transaction* platforms based on the internet of energy, artificial intelligence and big data analytics can prove to be an effective infrastructural solution. The grid operation could act as the platform owner at the initial phase and will see its role reduced to a minimum at later stages as the demand aggregator and the customer will get more freedom to perform transactions. They use the platform to independently file applications for load reduction; bidding is scheduled automatically; a demand aggregator is selected that meets the criteria; and technical conditions for contract implementation are determined. The key issues to be addressed as part of this model are associated with enabling access to the platform for generators, demand aggregators and consumers, the creation of adequate user interfaces, and defining the platform administrator mandate. The costs of creating and maintaining the operation of the platform could be covered either by a demand side management (DSM) surcharge imposed by the grid operator, or by a percentage charge deducted from the profits of demand aggregators. The size of the charge is to be established by law.

3.3 Asset management

Non-transaction platforms enable a transition to progressive risk-oriented strategies that imply continuous equipment condition monitoring and forecasting of possible outages in various nodes of technical systems. When using technology platforms as a knowledge-intensive solution enabling energy companies to radically rethink their production principles and those of relations with service vendors and customers, the key risk is posed by uncertainty as regards the right choice of assets that would make it possible to design a sufficient number of customized services in the long run without raising a large amount of additional investment [17]. The selection of assets that would enable an energy company to function in line with the platform model happens at the very initial stage of a production system overhaul. That is why it becomes ever more important to ensure close connections between the asset management strategy, the corporate strategy and the product strategy [18].

An example of platform utilization for asset management is the Cablewalker platform for robotic power line inspection that is being developed by the authors [19]. The platform enables the creation of an exact digital 3D model of a grid facility – a digital twin [20], containing information about the condition of every key element. A digital twin, in its turn, makes it possible to effectively manage the condition of grid assets that are hard-to-access for technical reasons. Test runs of the inspection method showed that up to 98% of defects on overhead transmission lines can be detected.

Another example is the organization of IoT-based tools for intellectual repairs and maintenance services on the basis of platforms [21]. Under the traditional approach, an energy company collects and stores information about the performance of mechanical and electrical equipment in a special database where the data is aggregated. The system is usually capable of calculating indirect parameters on the basis of technological process data, the current and retrospective conditions of mechanical and electrical equipment, and inspection results. The new approach envisages the platform-based globalization of various data influencing the equipment lifecycle. The system organizes the collection and processing of statistical data in real time, increasing the accuracy of the assessment of equipment condition and remaining lifetime and engaging remote users in the assessment process. With the application of the platform approach, the intellectual repairs and maintenance system acquires a number of essential system properties: scalability (capacity for integration with other systems; capacity for self-learning (the ability to change configuration settings during operation); self-organization (the ability to remain operational when new functions are added).

3.4 Arrangements of knowledge transfer between energy companies and universities

Platforms are self-organizing network structures whose objective is to generate new knowledge for tackling future industrial problems. In this context, *specialized science and education non-transaction platforms* are of particular demand.

The operation mechanism of such platforms (Fig. 3) ensures the circulation and multiplication of new knowledge that matches promising trends and developments in the global, national and industrial contexts. Their continuous monitoring and analysis make it possible to generate knowledge and convert it into concrete projects and research tasks. To address them, a portfolio of specific competences that are in demand by the market is needed [22].

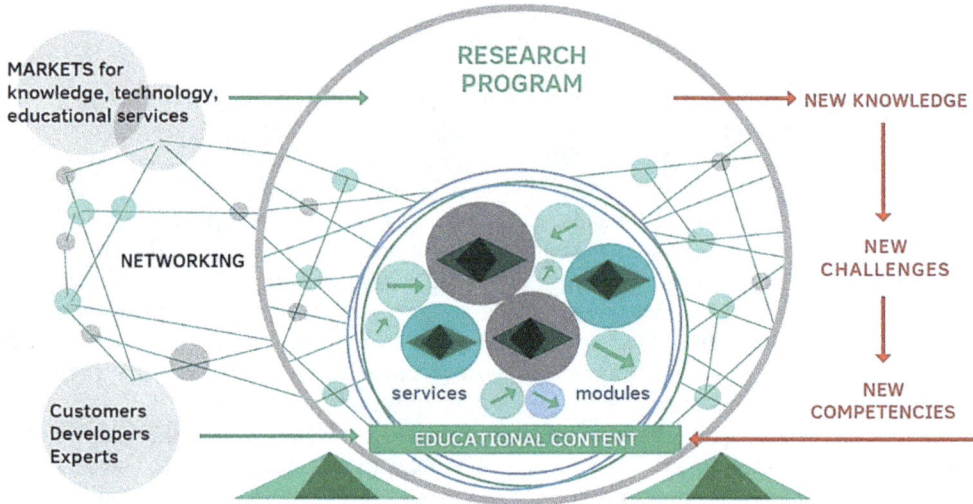

Figure 3: Conceptual image of science and education platform mechanism.

A knowledge base serves as the foundation for creating knowledge-intensive services across a science and education platform. The data base of the platform created by the authors on the premises of Ural Federal University contains over 50 proprietary course books, 400 articles and publications, making it possible to radically transform the instruction process by leaving theoretical aspects for self-study and freeing up time for focusing on problem areas and acquiring cross-disciplinary knowledge and competencies. The key services of the science and education platform are access to the knowledge base, publication collaborations [23], formation of breakthrough teams for unique projects, innovation tours, including online, of pioneering companies, design and adjustment of a specialist's personal development trajectory.

4 CONCLUSIONS

Technology platforms find growing adoption in manufacturing for the purpose of solving unique super-complex problems that are new and non-linear, while lacking a sufficient bank of knowledge and the essential analytical base. They can function as self-regulating consortiums or as scalable business eco-systems that produce services that flexibly adapt to market demand thanks to their unique technological architecture. Businesses using platforms for marketing are able to interact economically and directly with peers without any middlemen and additional capital and operation costs of financing market infrastructure.

In the electric power industry, platforms find demand in the following interconnected segments:

- smart grids, due to the radically growing intensity of interactions among grid companies and energy sellers and end users, the grid operator, dispatch organizations, ESCOs. This has become possible thanks to the adoption of smart metering technologies.

- in the "energy-as-a service" segment, energy companies are making a transition to a model of selling packages of knowledge-intensive services that are based on energy consumption management by means of adaptive information and communication technologies.
- the EV industry, following the emergence of new market agents – independent organizations or subsidiaries of energy companies that manage energy flows between energy consuming devices and the grid.

The authors show that platform-based instruments can be used in the electric power industry for addressing the following tasks: ensuring faster and more transparent energy transactions; price-dependent demand side response; asset management, knowledge transfer between energy companies and universities. All customers and suppliers of knowledge-intensive services are engaged in the implementation of these trends, which results in the arrival of new business models both within service companies that focus on customized high-tech industries, and energy companies, who see their economic behavior logic change towards a bigger share of services in the structure of their product offer instead of maximization of the sales of their standard core product (electricity).

It is worth noting that the implementation of the suggestions set out in this article requires further elaboration on certain debatable aspects that have to do with the creation of a regulatory framework for platform-based energy markets, cyber risk management throughout platform operation, integration of platforms with different technological architectures. These issues pose an interest for further research.

REFERENCES

[1]   Gitelman, L.D., Kozhevnikov, M.V. & Sandler, D.G., Technology platforms as a tool for solving complex innovation problems. *International Journal of Design and Nature and Ecodynamics*, **11**(4), pp. 584–592, 2016. DOI: 10.2495/DNE-V11-N4-584-592.
[2]   Gawer, A., Bridging differing perspectives on technological platforms: Toward an integrative framework. *Research Policy*, **7**(43), pp. 1239–1249, 2014.
[3]   Robertson, D. & Ulrich, K., Planning for product platforms. *Sloan Management Review*, **39**(4), pp. 19–31, 1998.
[4]   KPMG, The Changing Landscape of Disruptive Technologies. Tech Disruptors Outpace the Competition. https://assets.kpmg/content/dam/kpmg/pl/pdf/2018/06/pl-The-Changing-Landscape-of-Disruptive-Technologies-2018.pdf. Accessed on: 2 May 2022.
[5]   Lawrence, M., The next phase of the energy transformation: Platform thinking. https://www.renewableenergyworld.com/storage/the-next-phase-of-the-energy-transformation-platform-thinking/. Accessed on: 2 May 2022.
[6]   Weiller, C.M. & Pollitt, M.G., Platform markets and energy services. Working Paper, 2013. https://www.eprg.group.cam.ac.uk/wp-content/uploads/2013/12/1334-PDF.pdf. Accessed on: 2 May 2022.
[7]   Duch-Brown, N. & Rossetti, F., Digital platforms across the European regional energy markets. *Energy Policy*, **144**, 111612, 2020. DOI: 10.1016/j.enpol.2020.111612.
[8]   Martín-Lopo, M.M., Boal, J. & Sánchez-Miralles, Á., A literature review of IoT energy platforms aimed at end users. *Computer Networks*, **171**, 107101, 2020. DOI: 10.1016/j.comnet.2020.107101.

[9]   Shastitko, A.E. & Markova, O.A., An old friend is better than two new ones? Approaches to market research in the context of digital transformation for the antitrust laws enforcement. *Voprosy Ekonomiki*, **6**, pp. 37–55, 2020. (In Russian.) DOI: 10.32609/0042-8736-2020-6-37-55.

[10] Liang, T.-P., Lin, Y.-L. & Hou, H.-C., What drives consumers to adopt a sharing platform: An integrated model of value-based and transaction cost theories. *Information and Management*, **58**(4), 103471, 2021. DOI: 10.1016/j.im.2021.103471.

[11] Ceccagnoli, M., Forman, C., Huang, P. & Wu, D.J., Cocreation of value in a platform ecosystem: The case of enterprise software. *MIS Quarterly*, **36**(1), pp. 263–290, 2012.

[12] Parker, G. & Van Alstyne, M., Innovation, openness, and platform control. *Management Science*, **64**(7), pp. 3015–3032, 2018. DOI: 10.1287/mnsc.2017.2757.

[13] Afuah, A., Are network effects really all about size? The role of structure and conduct. *Strategic Management Journal*, **34**(3), pp. 257–273, 2013. DOI: 10.1002/smj.2013.

[14] Constantinides, P., Henfridsson, O. & Parker, G.G., Introduction: Platforms and infrastructures in the digital age. *Information Systems Research*, **29**(2), pp. 381–400, 2018. DOI: 10.1287/isre.2018.0794.

[15] Silvast, A., Williams, R., Hyysalo, S., Rommetveit, K. & Raab, C., Who 'uses' smart grids? The evolving nature of user representations in layered infrastructures. *Sustainability*, **10**, p. 3738, 2018. DOI: 10.3390/su10103738.

[16] Sui, H., Sun, Y. & Lee, W., A demand side management model based on advanced metering infrastructure. *2011 4th International Conference on Electric Utility Deregulation and Restructuring and Power Technologies (DRPT)*, pp. 1586–1589, 2011. DOI: 10.1109/DRPT.2011.5994150.

[17] Aracil, R., Pinto, E. & Ferre, M., Robots for live-power lines: Maintenance and inspection tasks. *IFAC Proceedings Volumes*, **35**(1), pp. 13–18, 2002.

[18] Gitelman, L.D., Kozhevnikov, M.V. & Gavrilova, T.B., Managing productive assets of energy companies in periods of crisis. *Challenges and Solutions in the Russian Energy Sector*, eds S. Syngellakis & C.A. Brebbia, Springer: Cham, pp. 55–61, 2018. DOI: 10.1007/978-3-319-75702-5_7.

[19] Gitelman, L.D., Kozhevnikov, M.V. & Kaplin, D.D., Asset management in grid companies using robotic fault diagnosis. *International Journal of Energy Production and Management*, **4**(3), pp. 230–243, 2019. DOI: 10.2495/EQ-V4-N3-230-243.

[20] Lamagna, M., Groppi, D., Nezhad, M.M. & Piras, G., A comprehensive review on Digital Twins for smart energy management system. *International Journal of Energy Production and Management*, **6**(4), pp. 323–334, 2021. DOI: 10.2495/EQ-V6-N4-323-334.

[21] Gitelman, L.D., Kozhevnikov, M.V., Chebotareva, G.S. & Kaimanova, O.A., Asset management of energy company based on risk-oriented strategy. *WIT Transactions on Ecology and the Environment*, vol. 246, WIT Press: Southampton and Boston, pp. 125–135, 2020. DOI: 10.2495/EPM200121.

[22] Gitelman, L.D., Kozhevnikov, M.V. & Ryzhuk, O.B., Advance management education for power-engineering and industry of the future. *Sustainability*, **11**(21), p. 5930, 2019. DOI: 10.3390/su11215930.

[23] Schiavon, M., Ragazzi, M., Magaril, E., Chashchin, M., Karaeva, A., Torretta, V. & Rada, E.C., Planning sustainability in higher education: three case studies. *WIT Transactions on Ecology and the Environment*, vol. 253, WIT Press: Southampton and Boston, pp. 99–110, 2021. DOI: 10.2495/SC210091.

REGULATIONS FOR EFFICIENCY ASSESSMENT OF INVESTMENT PROJECTS IN THE ENERGY SECTOR: BRIEF OVERVIEW AND COMPARATIVE ANALYSIS

ANZHELIKA KARAEVA[1], ELENA MAGARIL[1] & HUSSAIN H. AL-KAYIEM[2]
[1]Department of Environmental Economics, Ural Federal University, Russian Federation
[2]Mechanical Engineering Department, Universiti Teknologi PETRONAS, Malaysia

ABSTRACT

The policy of decarbonization and eco-modernization of the energy industry adopted by many countries implies a serious selection of investment projects primarily on the environmental pillar: projects aimed at reducing the negative impact on the environment or projects whose environmental damage from the implementation might be minimal, become priority. Therefore, the environmental impact assessment of investment projects became mandatory in the efficiency assessment process. Nowadays, there is no generally acclaimed methodology in the world for assessing the environmental and economic efficiency of investment projects: each country may have and apply its own requirements and methods or use recommendations developed and prepared by the international organizations. To ensure high efficiency of environmental and economic assessment, numerous normative tools, including recommendations and/or requirements for their implementation, are recommended that may vary significantly in different countries. The present study aims on comparative analysis of normative documents adopted in Russia, the USA and the EU to assess the environmental and economic efficiency of investment projects. In addition, the results of the analysis will help to identify the key differences in regulations and the main problems and difficulties of conducting an environmental and economic assessment of investment projects. Finally, this may enable further assessment of the possibilities of creating a unified global approach.

Keywords: efficiency assessment, investment project, regulations, energy efficiency, economic efficiency, environmental impact assessment, circular economy.

1 INTRODUCTION

The participant countries of the 26th UN Conference on Climate Change once again confirmed the importance of enhancing actions on mitigation of the anthropogenic impact on the environment [1]. At present, most governments are implementing the principles of the circular economy, including the use of renewable resources, recycling of secondary raw materials, development and use of renewable energy sources (RES) [3]–[5]. Therefore, in the context of the transition to a circular economy, any economic activity should demonstrate the prevention of negative environmental impact, rational use of resources and strive to minimize the generation of waste and other types of pollution.

The energy sector, as the world's largest environmental polluter [5], is expected to undergo significant changes in the upcoming years: eco-modernization of traditional energy facilities, dynamic development and introduction of renewable energy facilities, phase-out of coal use as fuel, etc. [6]–[8]. Large-scale transformations in the process of transition to a low-carbon economy are already stimulating the development and implementation of relevant investment projects in the energy sector: according to the International Energy Agency (IEA), the costs of research and development (R&D) in the energy sector worldwide increased from $26.9 billion in 2014 to $30.3 billion in 2019 at comparable prices for 2019 [9].

The inflows of investments and the current environmental agenda require a thorough selection of projects, taking into account both the economic and environmental pillars [10],

WIT Transactions on Ecology and the Environment, Vol 255, © 2022 WIT Press
www.witpress.com, ISSN 1743-3541 (on-line)
doi:10.2495/EPM220041

[11]. There is no unified recognized regulatory framework for efficiency assessment of investment projects in the energy sector at present.

Thereby, environmental impact assessment (EIA) procedure is mandatory within investment projects' selection and assessment. EIA allows making a decision on the implementation of a project not only by its economic parameters but also by a number of environmental characteristics [12]–[15].

It is important to note that EIA is just the one of the assessment stages that includes the analysis of the potential project's impact on the environment, the consideration of the long-term consequences and costs of environmental protection measures and the analysis of other environmental costs and energy efficiency of the project [11], [16]. A significant contribution to the development and improvement of approaches to EIA is made by the International Association for Impact Assessment (IAIA) that unites specialists in the field of assessing the social and environmental impact of programs or projects on the environment [17].

The purpose of this study is to review and compare key regulatory documents and standards for the environmental and economic assessment of investment projects in the United States, the European Union (EU) and Russia. It is expected that the results obtained will reveal the main features of regulatory documents, their content, approaches to evaluation and identify disadvantages and advantages.

2 MATERIALS AND METHODS

To analyse regulatory documents in the field of environmental and economic assessment of investment projects, the authors divided the considered documents into three groups:

1. Recommendations and/or guidance for general efficiency assessment of investment projects.
2. Recommendations and/or guidance for conducting Environmental Impact Assessment (EIA) of investment projects.
3. Recommendations and/or guidance for the energy efficiency assessment of investment projects (EEA).

The authors selected most significant documents in each of the above-mentioned groups used in the USA, the EU and Russia. To conduct a comparative analysis of regulatory documents on EIA and EEA, the authors also analysed the methodological recommendations of international organizations that are used in most countries in addition to the national regulatory documents. The list of documents is given in Table 1.

The list of criteria for comparative analysis of selected regulatory documents is presented in Table 2.

3 RESULTS AND DISCUSSION

3.1 Overview and comparative analysis of regulations for general efficiency assessment of investment projects

Among the fundamental international documents reflecting the methodology for efficiency assessment of investment projects are the "Evaluation Manual" developed by the United Nations Industrial Development Organization (UNIDO) and "Guide to Cost–Benefit Analysis of Investment Projects" developed by the European Commission (EC).

Methodological recommendations on investment project efficiency assessment, which are applied in Russia, are based on a document issued by UNIDO. Consequently, they have

Table 1: The list of considered regulatory documents and recommendations.

Name of the document	Organization/Country
Regulations for general efficiency assessment of investment projects	
Evaluation Manual	UNIDO
Guide to Cost–Benefit Analysis of Investment Projects	European Commission
Methodological Recommendations on Investment Project Efficiency Assessment	Russia
Regulations for environmental impact assessment	
Environmental Impact Assessment of Projects	European Union
National Environmental Policy Act (NEPA)	USA
On Approval of Requirements for Environmental Impact Assessment Materials	Russia
The World Bank OP 4.01 "Environmental Assessment"	The World Bank
Family of ISO 14000 "Environmental Management"	ISO
Regulations for energy efficiency assessment	
Guidebook for Energy Efficiency Evaluation, Measurement, and Verification: A Resource for State, Local, and Tribal Air and Energy Officials	USA
Commission Recommendations on Energy Efficiency First: from Principles to Practice. Guidelines and Examples for its Implementation in Decision-Making in the Energy Sector and Beyond	European Union
Methodological Recommendations for Evaluating the Effectiveness of the Implementation of Energy Saving and Energy Efficiency Measures in Industry	Russia
Family of ISO 50000 "Energy Management Systems"	ISO

Table 2: The list of criteria for comparative analysis of selected regulatory documents.

The group of documents	Criteria for comparative analysis
Regulations for general efficiency assessment of investment projects	1. Approaches to the assessment; 2. Assessment principles; 3. Step-by-step framework of the evaluation procedure; 4. Assessment recommendations by sector; 5. The presence of the EIA part; 6. Proposed methodology/ies for the EIA; 7. Indicators for the EIA; 8. Case-study.
Regulations for environmental impact assessment	1. Applicability to the investment project evaluation; 2. Principles of the EIA/EEA;
Regulations for energy efficiency assessment	3. Step-by-step framework of the EIA/EEA; 4. Indicators for the EIA/EEA; 5. The recommendations for EIA/EEA by sector; 6. Alternatives evaluation; 7. Recommendations for report preparation; 8. Case-study.

approximately the same structure, principles and approach to evaluation (cost-efficiency approach) [18], [19]. The significant differences between two considered documents are the presence of the case study and the system of indicators necessary for the assessment in the Russian version. Both differences could be described as the advantages of the document.

The recommendations developed by the EC [20] are based on the cost–benefit approach, contain a more detailed description of the assessment procedure for several sectors of the economy including energy, a case study for each sector and a separate methodology for calculating greenhouse gas emissions. In addition, the document contains a whole section of recommendations for EIA. In 2021, the EC prepared updated recommendations for efficiency assessment of investment projects [21] that are complementary to the considered document.

In all documents, much attention is given to the assessment of the economic and/or social effectiveness of the project. The assessment of its environmental performance is a separate assessment stage (EIA) that is managed by other regulations. The considered documents contain references to the need for its implementation, references to methods or generalized recommendations. For instance, the methodology for estimating greenhouse gas emissions is presented in the "Guide to Cost–Benefit Analysis of Investment Projects" in an abbreviated version, some recommendations for EIA are contained in the "Evaluation Manual" [18], [20].

The comparative analysis of the regulatory documents is presented in Table 3.

Thus, despite the similarity of individual elements of the considered regulations, there is a significant discrepancy in methodological approaches to assessing the impact of the project on the environment. To conduct an EIA, it is necessary to use additional tools and recommendations, or resort to the services of third-party

3.2 Overview and comparative analysis of regulations for general efficiency assessment of investment projects

3.2.1 National Environmental Policy Act (USA)

In 1970, the National Environmental Policy Act (NEPA) was signed in the USA that marked the beginning of activities in the field of EIA around the world. It marked the transition from

Table 3:   Comparative analysis of regulations for general efficiency assessment of investment projects.

Criteria	Evaluation Manual (UNIDO)	Guide to Cost–Benefit Analysis of Investment Projects (EU)	Methodological Recommendations on Investment Project Efficiency Assessment (Russia)
1. Approach to the assessment	Cost–effective approach	Cost–benefit approach	Cost-effective approach
2. Assessment principles	+	+	+
3. Step-by-step framework	+	+	+
4. Assessment by sector	–	+	–
5. The EIA part	+	+	–
6. Proposed methodology/ies for EIA	–	–	–
7. Indicators for the EIA	–	–	–
8. Case study	–	+	+

a reactive approach in environmental management to a preventive one, when possible anthropogenic consequences are assessed before making a final decision on the implementation of any activities.

The document obliges to conduct an environmental assessment of any state project and provide reports on the expected project's impact on the environment (environmental impact statement (EIS)) [22]. Due to the legislative peculiarities of the United States, the specifics of financing large projects and their implementation, almost all private projects are somehow connected with the state activities that makes the EIA procedure mandatory for them [27]. Typical EIS structure: introduction part: a statement of the purpose and need of the proposed action; description of the affected environment; range of alternatives to the proposed action; analysis of the environmental impacts of each of the possible alternatives.

The second part of the EIS involves an assessment of the impact of the project on the environment and includes an analysis of the impact of the planned activity on the quality of atmosphere, water resources, biological diversity of the territory where the activity is planned to be implemented.

NEPA does not contain information on recommended approaches to assessing the environmental effectiveness of the project, since the EIA with the subsequent preparation of the EIS is often carried out by specialized organizations or an invited group of specialists who possess the necessary methodological tools. However, the section "Analysis of the environmental impacts of each of the possible alternatives" provides a recommendation for using cost/schedule risk analysis.

On the basis of NEPA, regulatory documents were subsequently developed and adopted, fixing the need for EIA in the EU, Russia and other countries.

3.2.2 Environmental Impact Assessment of Projects (EU)

Two key directives are applied on the territory of the EU that enshrine the need for EIA of investment projects: the Strategic Environmental Assessment Directive (SEA) (Directive 2001/42/EC) [28] and the Environmental Impact Assessment Directive (EIA) (Directive 2011/92/EC) [29]. Requirements and regulations of Strategic Environmental Assessment Directive are used at the stage of selection and initial evaluation of investment projects on environmental and social aspects. Environmental Impact Assessment Directive contains requirements for the evaluation of already selected investment projects.

The key regulatory document in the field of EIA in the EU is the Environmental Impact Assessment of Projects manual, prepared and approved by the European Commission. The manual contains step-by-step instructions on preparing an Environmental Impact Assessment Report on an ongoing project or an already functioning enterprise and mandatory assessment elements such as climate change impact (mitigation and adaptation), risks of major accidents and disasters, biodiversity, use of natural resources. The document does not contain a specific procedure for conducting EIA, but some recommended methods are given [23].

The most frequently mentioned approaches are life cycle impact assessment (LCIA), multicriteria analysis and the use of special indicators to assess the impact of the project on the environment. The evaluation and preparation of the report can be carried out by a third-party organization that has a certain accreditation, or the project contractor independently, the choice of the evaluation methodology remains at the discretion of the project customer's specialists.

3.2.3 The Order of the Ministry of Natural Resources and Ecology of Russian Federation N 999 of December 1, 2020 "On approval of requirements for environmental impact assessment materials" (Russia)

At the legislative level, the need for an EIA procedure in Russia is enshrined in Article 32 of Federal Law N 7-FZ of 10.01.2002 (ed. of 12/30/2021) "On Environmental Protection" [30]. Requirements for the EIA procedure and materials since 2020 are contained in the Order of the Ministry of Natural Resources of the Russian Federation N 999 of December 1, 2020 "On approval of requirements for environmental impact assessment materials". The regulation includes the principles of EIA, requirements for materials and documentation on the results of the EIA, the need to consider alternative options for the implementation of the planned activity.

A feature of the Russian legislative act is the clause on mandatory public participation in the EIA procedure. The document does not contain a methodology for calculating the indicators required for the EIA.

3.2.4 The World Bank OP 4.01 "Environmental Assessment"

In 1989, the World Bank published recommendations for the EIA procedure. A distinctive feature of the recommendations is the division of all projects into three main categories

- Category A: Require a full EIA;
- Category B: Require only the environmental analysis;
- Category C: Not require an EIA.

Projects falling into categories A and B are required to assess the impact of the project on the quality of atmosphere, water resources, biodiversity and the territory where the project is planned to be implemented. At the same time, the recommendations do not contain a specific assessment methodology: it can be selected by the project organizers independently from officially approved methods on the territory of the state where the project is being implemented, or by invited specialists and organizations.

As in NEPA, the World Bank's EIA results report should include an assessment of alternative projects and solutions. At the same time, the methodology for calculating the ecological and economic efficiency of alternative projects and the project under consideration is not given. The document notes the need to compare economic results and socio-environmental consequences on the principle of "with the project" and "without the project".

The World Bank OP 4.01 "Environmental Assessment" is used in many countries around the world. In addition, the World Bank itself conducts EIA of socially significant projects. Based on this document, recommendations to the EIA European Bank for Reconstruction and Development (EBRD) were developed in 1992.

3.2.5 Family of ISO 14000 "Environmental management" issued by
       International Organization for Standardization

ISO standards are strategic tools in the form of recommendations and guidelines that help public and private companies to improve the efficiency of their activities and meet modern requirements in various fields.

The group of ISO 14000 includes 29 standards affecting various aspects of environmental management at enterprises [26]. As part of the efficiency assessment of investment projects, the following standards are applied:

- ISO/DIS 14005:2019 "Environmental management systems – Guidelines for a flexible approach to phased implementation";
- ISO 14044:2006 "Environmental management – Life cycle assessment (LCA) – Requirements and guidelines";
- ISO 14031:1999 "Environmental management. Environmental performance evaluation. Guidelines";
- ISO 14097:2021 "Greenhouse gas management and related activities" [26].

The application of the standards listed above makes it possible to conduct a deep EIA of an investment project, including at each stage of its life cycle. The potential impact of the project on the environment is often assessed only at the operational stage, while a negative impact can be exerted at the initial and elimination stages. The main advantage of the ISO 14000 family is the high level of consistency of all standards among themselves: for instance, the LCA procedure can be supplemented with indicators from ISO 14031 and ISO 14097:2021 that will significantly improve the quality of the assessment [26]. ISO standards can be applied to enterprises and projects implemented in various sectors of the economy regardless of their regional affiliation.

Table 4 presents a comparative analysis of the five reviewed documents in the field of EIA.

Guideline on Environmental Impact Assessment of Projects (EU) and The World Bank OP 4.01 "Environmental Assessment" showed the largest compliance with the selected criteria – six of eight.

NEPA is the first normative instrument stating the need for EIA, became the basis for the development of Guidance on Environmental Impact Assessment of Projects in the EU, OP 4.01 "Environmental Assessment" The World Bank and the Federal Law "On Environmental Protection" in Russia. In that regard, all documents have the same structure and a common set of EIA requirements.

The Family of ISO standards 14000 is the only group of international regulatory documents that enshrines approaches to assessing the environmental effectiveness of investment projects and provides a list of indicators. At the same time, ISO standards do not contain information about the need to compare alternative projects with each other and recommendations for compiling a report. This is due to the fact that the standards are advisory and can be applied worldwide.

The reviewed regulatory documents do not contain a single approved methodology for assessing the environmental effectiveness of projects. Most of the recommendations are

Table 4: Comparative analysis of regulations for EIA.

Criteria	National Environmental Policy Act (NEPA), (USA)	Guideline on EIA of Projects (EU)	On approval of Requirements for EIA Materials (Russia)	Family of ISO Standards 14000 "Environmental Management" (International)	The World Bank OP 4.01 "Environmental Assessment" (International)
1. Applicable to the investment project evaluation	+	+	+	+	+
2. Principles of EIA	+	+	+	+	+
3. Step-by-step methodology of EIA	+	+	+	+	+
4. Indicators to the environmental assessment	–	–	–	+	–
5. EIA by sector	–	–	–	–	–
6. Alternatives evaluation	+	+	+	–	+
7. Recommendations for report preparation	+	+	+	–	+
8. Case study	–	+	–	–	+

Table 5: Comparative analysis of regulations for EEA.

Criteria	Guidebook for Energy Efficiency Evaluation, Measurement, and Verification: A Resource for State, Local, and Tribal Air and Energy Officials (USA)	Commission Recommendations on Energy Efficiency First: From Principles to Practice (EU)	Methodological Recommendations for Evaluating the Effectiveness of the Implementation of Energy Saving and Energy Efficiency Measures in Industry	Family of ISO 50000 "Energy Management Systems"
1. Applicability to the investment project evaluation	+	+	+	+
2. Principles of EEA	+	+	+	+
3. Step-by-step methodology of EEA	+	+	+	+
4. Indicators to the EEA	+	–	+	+
5. EEA by sector	–	+	–	–
6. Alternatives evaluation	+	+	–	–
7. Recommendations for report preparation	+	–	–	+
8. Case study	+	–	–	–

based on a quantitative assessment of emissions of harmful substances into the atmosphere, the impact of the project on water resources and biodiversity of the territory.

Thus, there is currently no unified methodology or approved approaches to environmental efficiency assessment of investment projects. In most cases, EIA is reduced to a quantitative assessment of the negative impact on the environment: the volume of emissions/discharges of harmful substances, the volume of waste generation, the area of the territory used, etc. In this regard, comparing several investment projects of different scales is incorrect and can lead to errors in the decision-making process. In addition, in the USA and the EU, it is necessary to involve third-party specialists and organizations in conducting the assessment, while the right to choose the methodology remains with them, which in some cases may reduce the effectiveness of the assessment. For example, different approaches to assessing the impact of proposed activities on atmospheric air may lead to different results and conclusions regarding the effectiveness of the project, especially if projects from different countries are compared.

To carry out a deeper assessment, specialists and specialized organizations could use various approaches: LCIA, cost–benefit approach, cost-efficiency approach, multicriterial analysis, which are mentioned in the documents under consideration, but this is not a prerequisite for conducting an EIA.

3.3 Overview and comparative analysis of regulations for EEA

The energy efficiency assessment of an investment project can be characterized as an EIA component: the higher the energy efficiency of the project, the lower the use of natural resources (in particular, energy carriers) and, consequently, the negative impact on the environment [31].

Currently, there is no unified regulatory and legal apparatus for assessing the energy efficiency of investment projects. However, in the USA, EU countries, and Russia, there are a number of documents containing goals, principles, recommendations, and methods of energy efficiency assessment that can be resorted to as part of evaluating the effectiveness of investment projects.

3.3.1 Guidebook for Energy Efficiency Evaluation, Measurement, and Verification:
A Resource for State, Local, and Tribal Air and Energy Officials

This is a detailed guide to conducting an energy efficiency assessment at the national, regional level, at the level of an individual enterprise or investment project. The document contains requirements for evaluation, principles, approaches and indicators, case studies and served as the basis for the creation of similar methodological recommendations in many countries of the world.

3.3.2 Commission recommendations on Energy Efficiency First:
From principles to practice

Since 2012, Directive 2012/27/EU [36] has been in force in the EU, which prescribes the need to increase energy efficiency in the EU through the use of incentive measures to increase it in all sectors of the economy. The Directive sets out a general framework for improving energy efficiency and contains five key principles: (1) Energy, security, solidarity and trust; (2) A fully internal energy market; (3) Energy Efficiency first; (4) Transition to a long-lasting low-carbon society; (5) An Energy Union for Research, Innovation and Competitiveness. The directive covers existing enterprises in all sectors of the economy, public and private programs, social projects and investment projects.

3.3.3 Regulatory document "Commission Recommendations on Energy Efficiency First: From Principles to Practice"

Commission recommendations on Energy Efficiency First: from principles to practice" is an extensive methodological guide that includes not only the process of assessing energy efficiency, but also theoretical and practical issues of implementing and achieving targets.

3.3.4 Methodological recommendations for evaluating the effectiveness of the implementation of energy saving and energy efficiency measures in industry (the Order of the Ministry of Economic Development of the Russian Federation N 468 of July 29, 2019)

This provides the main approaches to assessing the effectiveness of energy saving and energy efficiency measures at the planning stage. Recommendations can be used for EEA of investment projects. It contains the theoretical basis and principles of evaluation, simplified evaluation procedure and indicators. It should be noted that Russia is currently implementing the state program of the Russian Federation "Energy Efficiency and Energy Development" [37], which is aimed at improving the energy efficiency of all sectors of the economy.

3.3.5 Family of ISO 50000 "Energy Management Systems"

From the point of view of the EEA evaluation of investment projects, the following ISO standards are of the greatest interest:

- ISO 50006:2014 "Energy management systems – Measuring energy performance using energy baselines (EnB) and energy performance indicators (EnPI) – General principles and guidance";
- ISO 50015:2014 "Energy management systems – Measurement and verification of energy performance of organizations – General principles and guidance";
- ISO 17743:2016 "Energy savings – Definition of a methodological framework applicable to calculation and reporting on energy savings".

By analogy with the ISO 14000 family of standards, ISO standards for energy efficiency assessment in general can be applied for both planned and ongoing projects. The standards contain general principles and approaches to energy efficiency assessment and are applied worldwide. The main advantage of ISO standards containing approaches to EIA and EEA is their high compatibility with each other.

Table 5 provides the results of the comparative analysis of regulations in EEA.

All the documents reviewed provide almost the same approach and principles for assessing the energy efficiency of investment projects. That fact indicates a sufficiently high degree of their uniformity, and the possibility of comparing alternative investment projects planned for implementation in different countries

The first analysed document is "Guidebook for Energy Efficiency Evaluation, Measurement, and Verification: A Resource for State, Local, and Tribal Air and The Energy Officials" developed by the EIA in the USA which meets all eight criteria of comparative analysis, while the document prepared in the EU is meeting six out of eight. The methodological recommendations for assessing energy efficiency in Russia meet four criteria, but contain key information: principles, approaches and indicators of EEA.

Thus, it can be concluded that the documentation in the field of EEA has the greatest international consistency in comparison with regulatory documents, official recommendations and standards for the overall assessment of the effectiveness of projects and EIA.

EIA and EEA are currently the components of the efficiency assessment procedure of the investment projects, i.e. documents from all three groups can be used simultaneously to evaluate one or a group of investment projects.

4 CONCLUSIONS

According to the given results, the authors note the feasibility of developing a unified methodological approach to conducting an environmental and economic assessment of the effectiveness of investment projects. At the present stage, the formation of such unified approach requires taking into account the basic principles of the circular economy. The reviewed regulatory documents mainly contain the principles and procedure for assessing the environmental effectiveness of projects, in the absence of methodological guidelines and indicators, or refer to additional regulations, adopted standards or guidelines. The creation of a single international document (for example, on the basis of UNIDO) with a list of approaches and guidelines for their use would allow the creation and consolidation of an international EIA standard.

In addition, in order to improve the effectiveness of the assessment and objectivity in the selection of projects, it is necessary to conduct a comprehensive assessment of the EIA, including the EEA, not only according to uniform methods, but also taking into account the specifics of different sectors of the economy. The ISO 14000 and ISO 50000 family of standards represent a systematized methodological toolkit that can be used in the process of EIA and EEA, respectively, and in the future can become the basis for creating unified approaches to the environmental and economic assessment of investment projects.

ACKNOWLEDGEMENT

The research was supported by the Russian Science Foundation grant No. 22-28-01740, https://rscf.ru/en/project/22-28-01740/.

REFERENCES

[1]   United Nations, COP26: Together for our planet. www.un.org/en/climatechange/cop26. Accessed on: 3 Feb. 2022.

[2]   Furlan, M. & Mariano, E., Guiding the nations through fair low-carbon economy cycles: A climate justice index proposal. *Ecological Indicators,* **125**, 107615, 2021. DOI: 10.1016/j.ecolind.2021.107615.

[3]   Su, C. & Urban, F., Circular economy for clean energy transitions: A new opportunity under the COVID-19 pandemic. *Applied Energy*, **289**, 116666, 2021. DOI: 10.1016/j.apenergy.2021.116666.

[4]   Dzikuć, M., Gorączkowska, J., Piwowar, A., Smoleński, R. & Kułyk, P., The analysis of the innovative potential of the energy sector and low-carbon development: A case study for Poland. *Energy Strategy Reviews,* **38**, 100769, 2021. DOI: 10.1016/j.esr.2021.100769.

[5]   Karaeva, A., Magaril, E., Torretta, V., Ragazzi, M. & Rada, E.C., Green energy development in an industrial region: A case-study of Sverdlovsk region. *Energy Report,* **7**, pp. 137–148, 2021. DOI: 10.1016/j.egyr.2021.08.101.

[6]   Bryant, T.S., Straker, K. & Wrigley, C., Designing our sustainable energy future: A shock doctrine for energy. *Energy Policy*, **147**, 111914, 2020. DOI: 10.1016/j.enpol.2020.111914.

[7]   Ahmad, T. & Zhang, D., A critical review of comparative global historical energy consumption and future demand: The story told so far. *Energy Report,* **6**, pp. 1973–1991, 2020. DOI: 10.1016/j.egyr.2020.07.020.

[8]   Xu, J.-H., Guo, J.-F., Peng, B., Nie, H. & Kemp, R., Energy growth sources and future energy-saving potentials in passenger transportation sector in China. *Energy,* **206**, 118142, 2020. DOI: 10.1016/j.energy.2020.118142.

[9]   International Energy Agency. www.iea.org/. Accessed on: 5 Feb. 2022.

[10]   Grubert, E. & Zacarias, M., Paradigm shifts for environmental assessment of decarbonizing energy systems: Emerging dominance of embodied impacts and design-oriented decision support needs. *Renewable and Sustainable Energy Reviews,* **159**, 112208, 2022. DOI: 10.1016/j.rser.2022.112208.

[11]   Karaeva, A., Magaril, E., Al-Kayiem H., Torretta, V. & Rada, E.C., Approaches to the assessment of ecological and economic efficiency of investment projects: Brief review and recommendations for improvements. *WIT Transactions on Ecology and Environment,* vol. 253, WIT Press: Southampton and Boston, pp. 515–525, 2021. DOI: 10.2495/SC210421.

[12]   Liu, Y., Xu, M., Ge, Y., Cui, C. & Xia, B. & Skitmore, M., Influences of environmental impact assessment on public acceptance of waste-to-energy incineration projects. *Journal of Cleaner Production,* **304**, 127062, 2021. DOI: 10.1016/j.jclepro.2021.127062.

[13]   Kulczycka, J. & Smol, M., Application LCA for eco-efficiency assessment of investment projects. *Acta Innovation,* **16**, pp. 29–38, 2015.

[14]   Caiado, R.G.G., Dias, R.F., Mattos, L.V., Quelhas, O.L.G. & Filho, W.L., Towards sustainable development through the perspective of eco-efficiency: A systematic literature review. *Journal of Cleaner Production,* **165**, pp. 890–904, 2017. DOI: 10.1016/j.jclepro.2017.07.166.

[15]   Ziyadin, S., Streltsova, E., Borodin, A., Kiseleva, N., Yakovenko, I. & Baimukhanbetova, E., Assessment of investment attractiveness of projects on the basis of environmental factors. *Sustainability,* **11**, p. 2544, 2019. DOI: 10.3390/su11092544.

[16]   Kicherer, A., Schaltegger, S., Tschochohei, H. & Pozo, B.F., Eco-efficiency: Combining life cycle assessment and life cycle costs via normalization. *International Journal of Life Cycle Assessment,* **12**, pp. 537–543, 2007. DOI: 10.3390/10.1065/lca2007.01.305.

[17]   International Association for Impact Assessment (IAIA). www.iaiasa.co.za/ Default.aspx. Accessed on: 5 Feb. 2022.

[18]   UNIDO Independent Evaluation Division, Evaluation manual. www.unido.org/sites/ default/files/files/2018-04/Evaluation%20Manual%20e-book.pdf. Accessed on: 5 Feb. 2022.

[19]   Methodological Recommendations on Investment Project Efficiency Assessment (approved by Ministry of Economy of the Russian Federation, Ministry of Finance of the Russian Federation, The State Committee of the Russian Federation on Construction, Architectural and Housing Policy on 21 Jul.1999 N VK 477. www.consultant.ru/document/cons_doc_LAW_28224/. Accessed on: 5 Feb. 2022.

[20]   Guide to Cost–Benefit Analysis of Investment Projects. www.ec.europa.eu/regional_ policy/sources/docgener/studies/pdf/cba_guide.pdf. Accessed on: 5 Feb. 2022.

[21]   Economic Appraisal Vademecum 2021–2027: General Principles and Sector Applications. www.ec.europa.eu/regional_policy/sources/docgener/guides/ vademecum_2127/vademecum_2127_en.pdf. Accessed on: 5 Feb. 2022.

[22]   National Environmental Policy Act (NEPA), United States Environmental Protection Agency. www.epa.gov/nepa. Accessed on: 7 Feb. 2022.

[23] Environmental Impact Assessment of Projects, Guidance on Scoping. www.ec.europa.eu/environment/eia/pdf/EIA_guidance_Scoping_final.pdf. Accessed on: 7 Feb. 2022.

[24] The Order of the Ministry of Natural Resources and Ecology of Russian Federation N 999 of December 1, 2020 "On approval of requirements for environmental impact assessment materials". www.docs.cntd.ru/document/573339130. Accessed on: 7 Feb. 2022.

[25] The World Bank OP 4.01 – Environmental Assessment. www.web.worldbank.org/archive/website01541/WEB/0__-2097.HTM. Accessed on: 7 Feb. 2022.

[26] ISO 14000 Family, Environmental management. www.iso.org/iso-14001-environmental-management.html. Accessed on: 7 Feb. 2022.

[27] Goldberg, D.M., A comparison of six environmental impact assessment regimes: The United States, The Czech Republic, Slovakia, The European Community, The World Bank, The European Bank for Reconstruction and Development. Center for International Environmental Law: California, USA, pp. 3–30, 1993. www.ciel.org/Publications/AComparisonof6EnvReg.pdf. Accessed on: 7 Feb. 2022.

[28] European SEA Directive 2001/42/EC. www.eur-lex.europa.eu/legal-content/EN/ALL/?uri=celex%3A32001L0042. Accessed on: 8 Feb. 2022.

[29] European IEA Directive 2011/92/EC. www.eur-lex.europa.eu/legal-content/EN/TXT/?uri=celex%3A32011L0092. Accessed on: 8 Feb. 2022.

[30] Federal Law N 7-FZ of 10.01.2002 (ed. of 12/30/2021) "On Environmental Protection". www.consultant.ru/document/cons_doc_LAW_34823/. Accessed on: 8 Feb. 2022.

[31] Đukić, M. & Zidar, M., Sustainability of investment projects with energy efficiency and non-energy efficiency costs: Case examples of public buildings. *Sustainability*, **13**, 5837, 2021. DOI: 10.3390/su13115837

[32] Guidebook for Energy Efficiency Evaluation, Measurement, and Verification: A Resource for State, Local, and Tribal Air and Energy Officials. www.epa.gov/sites/default/files/2015-08/documents/evaluation_guide.pdf. Accessed on: 8 Feb. 2022.

[33] Commission recommendations on Energy Efficiency First: from principles to practice. Guidelines and examples for its implementation in decision-making in the energy sector and beyond (EU). www.eur-lex.europa.eu/legal-content/EN/TXT/?uri=CELEX:32021H1749. Accessed on: 8 Feb. 2022.

[34] The Order of the Ministry of Economic Development of the Russian Federation N 468 of July 29, 2019 "Methodological recommendations for evaluating the effectiveness of the implementation of energy saving and energy efficiency measures in industry". www.docs.cntd.ru/document/560863760. Accessed on: 8 Feb. 2022.

[35] Family of ISO 50000. www.iso.org/files/live/sites/isoorg/files/store/en/PUB100400.pdf. Accessed on: 9 Feb. 2022.

[36] Directive 2012/27/EU. www.eur-lex.europa.eu/legal-content/en/TXT/?uri=celex:32012L0027. Accessed on: 9 Feb. 2022.

[37] The state program of the Russian Federation "Energy Efficiency and Energy Development". www.minenergo.gov.ru/node/323. Accessed on: 9 Feb. 2022.

ARTIFICIAL INTELLIGENCE-BASED DEMAND-SIDE RESPONSE MANAGEMENT OF RENEWABLE ENERGY

BAVLY HANNA, GUANDONG XU, XIANZHI WANG & JAHANGIR HOSSAIN
University of Technology Sydney, Australia

ABSTRACT

Renewable energy (RE) sources will aid in the decarbonization of the energy sector, which would assist in alleviating the negative consequences of climate change. However, using RE resources for hybrid power generation has two technological challenges, uncertainty and variability owing to RE features, making estimating generated power availability difficult. Artificial intelligence techniques have been used in a variety of applications in power systems, but demand-side response (DR) is just lately receiving major research interest. The DR is highlighted as one of the most promising ways of providing the electricity system with demand flexibility; as a result, many system operators believe that growing the scale and breadth of the DR programme is critical. There are many different sorts of demand reduction programmes, and the most common classification is dependent on who begins the demand reduction. There are two types of DR schemes: (1) price-based programmes and (2) incentive-based programmes.

Keywords: demand response, renewable energy, artificial intelligence, machine learning.

1 INTRODUCTION

Demand response (DR) has grown in importance in recent years in endorsing energy systems' stability and efficiency as a result of smart metering infrastructure and current improvements in information and communication technologies. It has the ability to manage inconsistencies in power demand (D) and supply (S) by managing elastic loads on the D-side [1]. In Energy Management System (EMS), a well-planned DR structure has major positive benefits for society. This includes increasing human comfort, promoting the use of renewable resources, lowering global energy consumption, and decreasing dependency on fuel resources linked with high carbon emissions [2].

Consumers can participate in DR programmes by lowering or adjusting their energy use during peak hours as a result of time-based tariffs or other financial incentives [3]. The most energy-intensive segments of electricity systems are residential and industrial customers (CUs). Residential and industrial sectors consumed 58% of aggregate global power usage in 2018 (22% and 36%, respectively), according to the U.S. Energy Information Administration [4].

The Home Energy Management System (HEMS) is described in the residential segment as the best system for delivering services for energy management to efficient management and monitoring of power usage, and effective generation, and storage in smart homes [5]. The primary manageable residential appliances are divided into three classifications: (1) thermostatically controlled appliances, (2) non-thermostatically controlled appliances, and (3) electrical energy storage.

Increasing urbanization, the integration of additional RE generation resources, and the outlook of electric vehicles introduce new challenges for the control of the power grid. The effects of these problems can already be observed in places like California, where rolling blackouts are becoming increasingly frequent, especially during the summer, when consumers need electricity the most and D is at its highest. Some of these problems can be tackled through additional capital investments to oversize the power grid at the transmission and distribution levels and create a buffer for D/S volatility. However, proper control

WIT Transactions on Ecology and the Environment, Vol 255, © 2022 WIT Press
www.witpress.com, ISSN 1743-3541 (on-line)
doi:10.2495/EPM220051

techniques can reduce the need for such investments significantly. In the U.S., buildings represent about 70% of the total power usage, and it is estimated that DR has the potential to reduce peaks of electricity D by roughly 7% to 27%, depending on the region [6].

The value of energy storage and DR can be assessed as a function of multiple factors, including the energy mix, how volatile the loads are, the costs of different energy generation and storage technologies, and what control systems are leveraged. Similarly, there are multiple tradeoffs between using centralized or distributed energy resources, which depend on the incremental unit costs of distributed energy resources and the incremental locational value of electricity [7]. Thus, there is a need for more simulation tools that can help in such kinds of assessments under diverse sets of conditions.

This paper is organized in five sections. Section 2 is a survey of literature on renewable energy demand-side response (DR). The applications of Artificial Intelligence (AI)/Machine Learning (ML) in the renewable energy sector are covered in Section 3. Section 4 discusses the problems of AI applications in DR management, followed by conclusions in Section 5.

2 RENEWABLE ENERGY DR

Price-based programmes and incentive-based programmes are the two categories of DR programmes. CUs in price-based schemes modify their energy use patterns as a response to changes in power prices. Unlike price-based programmes, control equipment is installed at the CU's location in incentive-based programmes. The Load Servicing Entities (LSEs) can use this technology to switch particular electric appliances at specific periods.

In price-based programmes, these members can save money by limiting their power use during peak hours or by receiving incentive payments in incentive-based programmes. Gathering DR resources to compete for ancillary services in the marketplaces can assist LSEs. DR, by actively participating in the power balance, adds to the overall dependability and stability of the power system. Furthermore, it aids in the avoidance or postponement of the building or implementation of distribution and transmission infrastructure [8]. By providing required information about individual consumers, such as their electricity consumption habits, load profiling plays a critical role in the formulation of lucrative D programmes.

2.1 Price-based DR programs

CUs alter their power use in accordance with the price set by LSE in price-based schemes, such as (1) Time Of Use (TOU), (2) Critical Peak Pricing (CPP), and (3) Real Time Pricing (RTP). Only the overall electrical use of a day, month, or even more extended period is sent to the LSE without smart meters. Consumers with similar overall consumption but varying peak consumption are charged the same amount. Load profiling may be done with the use of smart meters to help the LSE establish time-variant pricing in order to optimize profit and increase DR.

Creating an optimization model and solving it based on load profiling is the basic approach to pricing design. A three-stage technique for TOU design was developed by Mahmoudi-Kohan et al. [9]. Various rates are optimized for CUs belonging to different clusters independently after determining the CUs' eagerness to purchase power. A simulation with 300 clients revealed that clustering with a lower CDI might result in higher profits. Conditional Value at Risk (CVaR) is applied by Mahmoudi-Kohan et al. [10]. The stochastic programming model was used using the same clustering approach and acceptance function as shown by Mahmoudi-Kohan et al. [9].

Mahmoudi-Kohan et al. show that load profiling was performed to discover CUs with comparable peaks in the overall curve [11]. CUs with high elasticity are given priority in a cost reduction model. The limitation states that the overall load decrease must equal or exceed the retailer's power shortfall. Panapakidis et al. used load profiling to extract typical load curves and then determined the pricing for each standard curve [12]. For an industrial CU, a comparison was made between yearly load profiling and seasonal load profiling. The findings revealed that taking into account the impact of seasonal pricing variations in the pool market might boost earnings. Maigha and Crow offered load profiling-based optimum TOU structures instead of pricing design [13]. A performance parameter for clustering algorithms is the granularity of the cluster's sensitivity, that relates to a scenario in which clusters comprise fewer than two hours and are scattered throughout the day. A case study confirmed that among the examined strategies, Gaussian mixture models performed the best.

2.2 Incentive-based DR programs

In incentive-based schemes, the residential CU has a contract with the programme administrator (such as an "aggregator" or "service provider") under which the programme administrator may be authorized to execute various control measures aimed at lowering power prices. The following DR programmes are often accessible in this category: (1) Direct Load Control (DLC), (2) Interruptible Tariffs, (3) Demand Bidding Programs, and (4) Emergency Programs [14]. Since it invades the CU's privacy, the DLC technique is deemed intrusive.

The administrator has authority over the operation of the client's appliance until the CU gets the agreed-upon payment in DLC. Interruptible tariffs are available to both "industrial" and "residential" consumers, and they basically provide multiple pricing layers according to the agreement between the energy provider and the client. The quantity of energy utilized is not reduced by load interruption, but it does shift load operation to off-peak hours [15]. CUs can take part in the electricity trading market by proposing to adjust their patterns of consumption, reschedule their loads, or reduce their usage through D-bidding. During times of high D or when the grid is disrupted by unanticipated occurrences, emergency programmes are implemented. Participants in these programmes limit their consumption to relieve grid stress in emergency circumstances, and they are compensated with compensation depending on the required amount of load reduction [16].

Furthermore, the temperature sensitivity of home power was the subject of Albert and Rajagopal's research while segmenting electrical consumers [17]. A novel probabilistic graphical model was used to simulate each CU's thermal regimes, with each concealed state representing the usage of heating or cooling equipment. Effective duration and effective thermal response sensitivity were used to classify the consumers. Because of the large potential for the usage of heating and/or cooling equipment, the thermal profile of consumers is critical for DR.

3 AI AND ML APPROACHES

The capacity of a smart grid with integrated RE to decrease CU suffering while maintaining fair power costs is critical to its future success. In order to balance the energy generation and delivery options, and AI/ML-enabled DR program may incorporate CUs into the decision-making process. The ever-increasing need for energy has dramatically widened the gap between D/S during rush hours, resulting in a huge increase in the financial worth of grid-connected electricity.

3.1 Artificial neural network based DR management

Artificial Neural Network (ANNs) are computer models that are based on the biological nervous system. It is divided into two categories: (1) "single hidden layer ANN" and (2) "deep learning". In the case of DR, a "single hidden layer ANN" is used to classify consumers, loads, and prices, whereas deep learning aids in predicting consumer reaction behavior, controlling household equipment, clustering consumers, and so on. Table 1 summarizes recent developments.

Table 1: ANN based DR management.

Reference	Year	Technique(s)	Objective(s)
Shirsat and Tang [18]	2021	Linear regression, MD-RNN	Determine the CU's heat sensitivity
Ruan et al. [19]	2020	NNLMS model	Speed up the distributed DR mechanism
Hafeez et al. [20]	2020	ANN-based forecast engine, a DA-GmEDE-based HEMC	Lower power bills, relieve the PAR
Lu and Hong [21]	2019	RL, DNN	Stimulate D-side involvement, boost SP and CU profitability, and enhance system dependability
Lu et al. [22]	2019	ANN, multi-agent RL	Determine the best judgments for diverse appliances

Shirsat and Tang proposed a system for determining distinct CUs' consumption reduction potential and generating mixed distributions to assess their reduction capabilities [18]. They utilize these distributions to create simulations for a stochastic knapsack problem with risk aversion. The stochastic CU selection issue has been addressed for the first-time employing mixture distributions and CVaR as an empirical risk metric. To predict the load decrease during a DR event, their suggested model relies on the sensitivity of consumers' consumption to external temperature.

The delayed convergence of existing distributed DR systems makes it difficult to develop dependable smart grid applications. To deal with this problem, Ruan et al. present a novel distributed procedure, the neural-network-based Lagrange multiplier selection, that significantly reduces iterations while avoiding oscillation [19]. The major enhancement is in an LSE forecast technique, which uses a specifically developed neural network (NN) to record consumers' pricing reaction attributes.

In the smart grid, Hafeez et al. presented a unique framework for effective HEMS to lower power bills, relieve the Peak-to-Average Ratio (PAR), and get the desirable balance between energy costs and user discomfort [20]. For efficient energy management, the forecasting software anticipates price-based DR signals and usage of energy trends. The home energy management controller (HEMC) timetables home equipment based on the anticipated energy usage patterns and pricing signals. For performance validation, the suggested day-ahead grey wolf modified enhanced differential evolution algorithm (DA-GmEDE) based approach is compared to two benchmark strategies: day-ahead genetic algorithm and day-ahead game-theory.

Lu and Hong developed a real-time incentive-based DR algorithm with Reinforcement Learning (RL) and Deep Neural Network (DNN) for smart grid systems, with the goal of assisting the Service Provider (SP) in acquiring energy resources from its various CUs to balance energy variations and improve grid dependability [21]. To address future uncertainties, the SP can only access the price from the wholesale power market and energy D from its CUs for the present hour due to the intrinsic nature of real-time electricity markets.

Lu et al. suggested an hour-ahead DR method for HEMS, based on AI, with the goal of reducing the user's energy bill and level of discomfort [22]. A stable price forecasting model based on ANN is proposed to overcome future pricing uncertainty. Multi-agent RL is used in conjunction with anticipated pricing to make the best selections for various appliances.

3.2 ML based DR management

ML is a set of techniques for identifying patterns in empirical data and turning them into useable models. supervised learning, unsupervised learning, and RL are the three basic forms of ML used in DR algorithms, as shown in Table 2.

Table 2: ML based DR management.

Reference	Year	Technique(s)	Objective(s)
Yang et al. [23]	2022	KELM, APVMD, CSCA	Power price forecasting tool
Pallonetto et al. [24]	2022	LSTMs, SVM	Load data forecasting comparison
Wicaksono et al. [25]	2021	DNN, LSTM, CNN, Hybrid	Estimating DR program's dynamic electricity pricing
Uimonen et al. [26]	2020	RF, NNMs	Power curtailment at a low cost
Pallonetto et al. [27]	2019	ML approaches for data modeling and optimization algorithm	Deployment of DR techniques

Yang et al. created an enhanced electricity price forecasting model using adaptive data pretreatment, sophisticated optimization, kernel-based model, and optimum model selection technique [23]. An Adaptive Parameter-Based Variational Mode Decomposition (APVMD) method is developed to achieve appropriate data preprocessing outcomes. The Chaotic Sine Cosine Algorithm (CSCA) is used to design and implement a leave-one-out optimization strategy for developing effective Kernel-Based Extreme Learning Machine Models (KELM) and APVMD.

Pallonetto et al. assess and evaluate the two most often used short-term load forecasting methods [24]. They teach the basics of Long Short-Term Memory Networks (LSTMs) and Support Vector Machines (SVM), as well as the typical techniques of short-term load forecasting. Preprocessing of data and feature selection are then performed based on the features of the electrical load dataset. One-hour ahead load forecasting and Peak and valley load forecasting one day ahead are done using the LSTMs and SVM models.

Wicaksono et al. create a system that uses pricing and incentive-based DR programmes to engage manufacturing power CUs [25]. Instead of centralized data integration, the system uses data from heterogeneous systems on both the D/S sides, which are linked by semantic middleware. The semantic middleware uses an ontology as its integrated information model.

ML algorithms are being developed to anticipate the power provided by RE sources as well as the electricity consumed by manufacturing users based on their operations.

Uimonen et al. offered a solution to the problem of selecting acceptable escalators from a vast pool in order to meet the aim of power curtailment at a low cost, and they highlighted the escalator attributes that make the best DR candidates [26]. They examine four different calculating methods that differ in computation speed and accuracy. The primary answer is the simulation-based model that was previously created and improved. The random forest (RF) and neural network models (NNMs) give a solution based on the simulation-based model's output, with the goal of increasing computation speed.

Pallonetto et al. evaluated the performance of control algorithms in the residential sector for the deployment of DR techniques [27]. A calibrated building simulation model was constructed and used to evaluate the effectiveness of DR techniques in combination with thermal zone management under various time-of-use power tariffs. Two DR algorithms were used to manage an integrated heat pump and thermal storage system, one based on a rule-based approach and the other on a predictive-based ML method. A common DR pricing scheme was used to compare the two algorithms.

3.3 Nature-inspired algorithm based DR management

In search processes, nature-inspired algorithms (NIA) are used to forecast the sequence of activities required to attain the stated goals. Evolutionary algorithms, biological swarms, and physical processes are all frequent DR methods, as shown in Table 3.

Table 3: NIA based DR management.

Reference	Year	Technique(s)	Objective(s)
Singh et al. [28]	2021	Black Widow Optimization, Technique for Order of Preference by Similarity to Ideal Solutions	Improves the load factor and system dependability
Bui et al. [29]	2020	SI, ABC	Optimize power costs of smart HEMS
Makhadmeh et al. [30]	2019	GA, GWO	Minimize the electricity bill and PAR
Ullah and Hussain [31]	2019	GA, MFO, TG-MFO	Lower energy costs
Silva and Han [32]	2019	ACO	Overall cost reduction

To overcome the uncertainty associated with solar and wind power output, Singh et al. used a stochastic-based scenario development and reduction strategy [28]. Unlike other techniques, the flexible load responsive model is developed for each DR programme in order to quantify the sensitivity of consumer engagement. TOU, CPP, RTP, and a mix of both TOU and CPP are used to modify predicted load D. The suggested problem is analyzed on a three-feeder microgrid (MG) test system, and Black Widow Optimization is used to find the best scheduling configuration for DR programmes.

By incorporating the notion of swarm intelligence (SI) into connected devices, Bui et al. offer a computational intelligence model for IoT applications [29]. Decentralized management of smart HEMS is taken into account, in which linked appliances make individual decisions for optimizing power costs of smart HEMS by exchanging information with one another. They are divided into two primary categories: (a) they propose a framework for decentralized management in smart HEMS; and (b) the artificial bee colony (ABC) algorithm, a typical SI technique, has been applied to connected appliances in terms of communication and collaboration with one another to optimize the EMS' performance.

For the Power Scheduling Problem (PSP), Makhadmeh et al. used the Multi-Objective Grey Wolf (GWO) optimizer [30]. PSP is handled by setting household equipment to a certain time horizon to reduce power bills and PAR while also improving user comfort. To produce an optimal schedule, the multi-objective function is formalized and used in GWO. The suggested multi-objective GWO is evaluated using seven consumption profiles and seven real-time energy prices with distinct features. The suggested algorithm's performance is evaluated using three criteria: electricity bill, PAR, and user comfort level.

Ullah and Hussain suggested two bio-inspired heuristic algorithms for an EMS in smart homes and buildings: the Moth-Flame Optimization (MFO) method and the Genetic Algorithm (GA) [31]. The performance of these devices in terms of energy cost reduction, PAR minimization, and end-user discomfort minimization is examined. Then, to meet the aforementioned goals, a hybrid version of GA and MFO called Time-constrained Genetic-Moth-Flame Optimization (TG-MFO) is presented. To provide optimal end-user comfort, TG-MFO not only combines GA, and MFO, but also integrates time limitations for each appliance. In the literature, many energy optimization strategies have been presented.

Appliance scheduling using heuristic algorithms is being studied as a possible option for managing the energy D/S gap during peak hours. However, because of the potential for early convergence, the validity of Ant Colony Optimization (ACO) based scheduling has been questioned. As a result, Silva and Han suggested a mutation operator integrated ACO scheduling method with pre-defined consumption limits to reduce energy costs and waiting time while addressing ACO's contested shortcoming [32]. The comparative study verifies the suggested work's superiority in terms of cost reduction, peak load reduction, waiting time reduction, and PAR reduction, indicating its potential to become a mainstream solution for D-side management challenges.

3.4 Multi-agent based DR management

A multi-agent system (MAS) is a "computerized system" that is composed of multiple intelligent agents that communicate with each other. Multiple interacting intelligent agents can be used in DR projects to enable successful planning, choices, and methods for RE resources. "Coalitional game theory", "mechanism design", and "automated negotiation" are the three subsections of MAS. They also help with DR programmes, as shown in Table 4.

A three-layer MAS optimization model including Distributed Management System (DMS) agent, MG Central Controller (MGCC) agent and MG Controllable Element (MGCE) agent are built by Li et al. [33]. Then the DR power and heat load mechanism is constructed, with the real-time production of new energy generation and the Energy Storage System (ESS) as optimization objects and the operational cost, environmental cost, and wind and solar abandonment cost as optimization targets. They suggest an improved particle swarm optimization approach based on an adaptive-weight and chaotic search to solve this problem. Finally, three scenarios are presented to demonstrate that the ESS and DR may lower the cost of MGs while also encouraging new energy use, as well as the superiority of the enhanced algorithm.

Vázquez-Canteli et al. introduced CityLearn, which is an OpenAI Gym environment, and a simulated framework for the implementation of RL for DRM and urban energy management [34]. CityLearn guarantees that, at any time, the heating and cooling energy D

Table 4: Multi-agent based DR management.

Reference	Year	Technique(s)	Objective(s)
Li et al. [33]	2020	DMS agent, MGCC agent, MGCE agent	Encompass the DR of electrical and heat load, as well as the ESS
Vázquez-Canteli et al. [34]	2020	Multi-agent, single-agent RL algorithms	Customize the incentive function and select reward modes (central-agent or multi-agent)
Golmohamadi et al. [35]	2019	DRP, IDRA, RDRA	Provide up/down-regulation for the power system in case of a deficiency or excess of generation
Li et al. [36]	2018	Stackelberg–Cournot game model with two stages	Integrate RE and DR into the wholesale power market
Leo et al. [37]	2018	JADE in Eclipse IDE	Stabilize and optimize the MG

of the building are satisfied regardless of the actions of the controller. The actions of the RL controller are automatically overridden to satisfy such constraints of thermal energy D. This allows the controllers to focus on shaping the curve of electricity consumption without running the risk of interfering with the comfort of the occupants or the desired temperatures.

Golmohamadi et al. provide a unique market-based strategy for integrating the flexibility potential of diverse, responsive CUs, such as the residential and industrial sectors, into a power system with substantial intermittent power penetration [35]. The ultimate goal was to offer up/down control for the power system in the event of a generating deficiency or surplus. The complex challenge was divided into a multi-agent framework to achieve the goal. As a result, three types of agents were studied: DR Provider (DRP), Industrial DR Aggregators (IDRA), and Residential DR Aggregators (RDRA). The time-oriented DR programme was developed to optimize the overall cost of energy and regulation by allowing the DRP to trade DR opportunities in three consecutive floors of the electricity market, namely the day-ahead, adjustment, and balancing markets, in order to ensure power system flexibility. Instead of subsidizing D-side flexibility, the DRAs might exchange DR values in a competitive framework based on D bids.

Li et al. propose a novel RES and DR programme integration architecture to increase energy efficiency and system resilience [36]. They propose a two-stage Stackelberg–Cournot game model to describe energy trading behaviors among power utilities, RES-based MGs, and DR players. In the suggested paradigm, the power utility is the leader, while RES-based MGs and DR participants coordinated via an aggregator are the followers. Furthermore, using a risk-controlled game model, a CVaR assessment is used to quantify the intermittency of RES and the uncertainty of DR, which might lead to a more dependable energy trading strategy for both the forwards and spot markets. Finally, they provide a computational approach that accelerates the optimal reaction dynamics on the follower side.

Leo et al. create a simulation model for dynamic energy management that takes into account the intermittent nature of solar power, randomness of load, dynamic grid pricing, and variation of critical loads, and chooses the best possible action every hour to stabilize and optimize the micro-grid using Java Agent Development Environment (JADE) in Eclipse IDE [37]. Additionally, environmental factors are detected by an Arduino Mega microcontroller and sent to MAS agents. MAS improves responsiveness, stability, adaptability, and fault tolerance, resulting in increased operational efficiency and cost and environmental savings.

4 DISCUSSION

DR indicates changes in energy consumption patterns, such as through financial incentives or improved consumption optimization, in order to better match the power supply in power systems [1]. It has complicated impacts on integration costs in general, making it significant for various cost components in addition to profile costs [2]. DR can reduce profile costs in the long run by decreasing peak D and boosting capital utilization while deferring the need for network improvements, hence affecting grid-related integration costs. In the short term, it has an impact on power markets, which might affect balancing costs.

Rocha et al. describe a novel energy planning methodology based on AI methods for smart homes [38]. This study takes into account power price variations, equipment priority, operational cycles, and a battery bank to anticipate distributed generation. When smart houses with and without distributed generation and battery banks were examined, the method's efficiency revealed a 51.4% cost savings.

The majority of the research on DR programmes for residential CUs focuses on developing a model of household loads that may be used to identify an electric usage pattern. This is accomplished by using either a grid-oriented approach, which models end-user consumption as a whole in terms of general characteristics such as gross domestic product and unemployment rate, or a scenario-oriented approach, by Appling a bottom-up technique, in which the load profile is created by aggregating the electric consumption of numerous domestic appliances or a variety of families [39]. Such research uses a simulation-based optimization analysis to estimate the advantages of providing disaster recovery services to residential clients [40].

According to several stakeholders, the primary challenges to DR are inadequate programme design and low CU engagement. Because better programme design may enhance client engagement, there is a strong link between these two obstacles. The bulk of currently used DR systems, which are based on highly centralized control ideas, need the collection and processing of a considerable quantity of local data from a central location. This demonstrates a great deal of complexity at the central coordinating point, which has an impact on the scalability of such DR methods. As a result, in all DR deployments, the majority of the controllable D concerns large commercial or industrial clients who fail to include a significant number of modest residential CUs.

We discovered that RL could adapt to its surroundings and acquire CU preferences through a feedback control loop throughout the review process, which appears promising for MG planning. When a large amount of data is available, RL algorithms can be effective, and system control is based on real-time judgments.

Modeling and computation processes are growing significantly more challenging as issue sizes become larger. Market operators must hedge against the more complicated structure in today's restructured energy markets by providing market players with adapting tools to the new market structure. To deal with such a challenge, MAS is a way to break down a large problem into smaller pieces. In this method, different agents may simply replicate the market model, which can then be expanded by additional entities.

MAS is particularly effective at solving complicated issues. A wide range of applications employs MAS. Gazafroudi et al. describe a MAS for the intelligent use of power in a smart home, resulting in increased energy efficiency [41]. MAS method has been widely embraced in the bottom-up approach because of its scalability and capacity to mimic the stochastic nature of household consumption as well as the dynamic interactions between residences and the grid. The power system literature has various MAS-based applications, including (1) electricity market, (2) voltage regulation, (3) load restoration, (4) load shedding, and (5) the smart grid area.

Pallonetto et al. examined the implementation of D-side management methods in the residential sector using a rule-based and predictive ML algorithm [42]. The rule-based system saved 20.5% on electrical end-use expenditures compared to the baseline scenario, while the predictive algorithm saved 41.8%. For utility generating costs, both strategies are saved in the same range.

5 CONCLUSIONS

Through price modifications or incentives, a DR programme encourages end-users to modify their power consumption habits to match RE sources' availability. The adoption of DR systems is a steady but sluggish trend aimed at maximizing the usage of RE in residential homes and many industries, including manufacturing. We analyzed four AI approaches, namely (1) ANNs, (2) ML, (3) NIAs, and (4) MAS, which have been utilized in DR programs to support RE dissemination. We discovered a DR program based on ANN that helped to reduce load shedding, enhance the PAR, engage in energy management schemes, and optimize load D, among other things. ML techniques were used for clustering, to reduce peak consumption, and for optimal bidding strategy to reduce the uncertainty of consumer's D and flexibility. The NIA was employed to optimize the scheduling of distributed energy resources, provide incentive-based DR management, control smart devices, and benefit the aggregator. For optimization purposes, a multi-agent-based DR program enabled bidding strategy, energy management, and power trading.

REFERENCES

[1] Yu, M., Lu, R. & Hong, S.H., A real-time decision model for industrial load management in a smart grid. *Appl. Energy*, **183**, pp. 1488–1497, 2016.
[2] Li, Y.-C. & Hong, S.H., Real-time demand bidding for energy management in discrete manufacturing facilities. *IEEE Trans. Ind. Electron.*, **64**(1), pp. 739–749, 2017.
[3] Department of Energy (DOE), 2022. https://www.energy.gov/.
[4] U.S. Energy Information Administration, 2022. https://www.eia.gov/.
[5] Zhou, B. et al., Smart home energy management systems: Concept, configurations, and scheduling strategies. *Renew Sustain Energy Rev*, **61**, pp. 30–40, 2016.
[6] Gils, H.C., Assessment of the theoretical demand response potential in Europe. *Energy*, **67**, pp. 1–18, 2014. DOI: 10.1016/j.energy.2014.02.019.
[7] Burger, S.P., Jenkins, J.D., Huntington, S.C. & Pérez-Arriaga, I.J., Why distributed? A critical review of the tradeoffs between centralized and decentralized resources. *IEEE Power and Energy Magazine*, **17**(2), pp. 16–24, 2019.
[8] Connell, N.O., Pinson, P., Madsen, H. & Malley, M.O., Benefits and challenges of electrical demand response: A critical review. *Renewable and Sustainable Energy Reviews*, **39**, pp. 686–699, 2014.
[9] Mahmoudi-Kohan, N., Parsa Moghaddam, M., Sheikh-El-Eslami, M.K. & Shayesteh, E., A three-stage strategy for optimal price offering by a retailer based on clustering techniques. *International Journal of Electrical Power and Energy Systems*, **32**(10), pp. 1135–1142, 2010.
[10] Mahmoudi-Kohan, N., Parsa Moghaddam, M. & Sheikh-El-Eslami, M.K., An annual framework for clustering-based pricing for an electricity retailer. *Electric Power Systems Research*, **80**(9), pp. 1042–1048, 2010.

[11] Mahmoudi-Kohan, N., Eghbal, M. & Moghaddam, M.P., Customer recognition-based demand response implementation by an electricity retailer. *21st Australasian Universities Power Engineering Conference (AUPEC)*, IEEE, pp. 1–6, 2011.

[12] Panapakidis, I.P., Simoglou, C.K., Alexiadis, M.C. & Papagiannis, G.K., Determination of the optimal electricity selling price of a retailer via load profiling. *47th International Universities Power Engineering Conference (UPEC)*, IEEE, pp. 1–6, 2012.

[13] Maigha & Crow, M.L., Clustering-based methodology for optimal residential time of use design structure. *North American Power Symposium (NAPS)*, IEEE, pp. 1–6, 2014.

[14] Haider, H.T., See, O.H. & Elmenreich, W.A., Review of residential demand response of smart grid. *Renew. Sustain. Energy Rev.*, **59**, pp. 166–178, 2016.

[15] Kostková, K., Omelina, L., Kyčina, P. & Jamrich, P., An introduction to load management. *Electr. Power Syst. Res.*, **95**, pp. 184–191, 2013.

[16] Chen, C., *Demand Response: An Enabling Technology to Achieve Energy Efficiency in a Smart Grid. Application of Smart Grid Technologies*, Elsevier: Amsterdam, pp. 143–171, 2018.

[17] Albert, A. & Rajagopal, R., Thermal profiling of residential energy use. *IEEE Transactions on Power System*, **30**(2), pp. 602–611, 2015.

[18] Shirsat, A. & Tang, W., Quantifying residential demand response potential using a mixture density recurrent neural network. *International Journal of Electrical Power and Energy Systems*, **130**, 106853, 2021.

[19] Ruan, G., Zhong, H., Wang, J., Xia, Q. & Kang, C., Neural-network-based Lagrange multiplier selection for distributed demand response in smart grid. *Applied Energy*, **264**, 2020. DOI: 10.1016/j.apenergy.2020.114636.

[20] Hafeez, G. et al., An innovative optimization strategy for efficient energy management with day-ahead demand response signal and energy consumption forecasting in smart grid using artificial neural network. *IEEE Access*, **8**, pp. 84415-84433, 2020. DOI: 10.1109/ACCESS.2020.2989316.

[21] Lu, R. & Hong, S.H., Incentive-based demand response for smart grid with reinforcement learning and deep neural network. *Applied Energy*, **236**(C), pp. 937–949, 2019.

[22] Lu, R., Hong, S. & Yu, M., Demand response for home energy management using reinforcement learning and artificial neural network. *IEEE Transactions on Smart Grid*, **10**(6), pp. 6629-6639, 2019. DOI: 10.1109/TSG.2019.2909266.

[23] Yang, W., Sun, S., Hao, Y. & Wang, S., A novel machine learning-based electricity price forecasting model based on optimal model selection strategy. *Energy*, **238**, 121989, 2022. DOI: 10.1016/j.energy.2021.121989.

[24] Pallonetto, F., Mangina, E. & Jin, C., Forecast electricity demand in commercial building with machine learning models to enable demand response programs. *Energy and AI*, **7**, 2021. DOI: 10.1016/j.egyai.2021.100121.

[25] Wicaksono, H., Boroukhian, T. & Bashyal, A., A demand-response system for sustainable manufacturing using linked data and machine learning. *Dynamics in Logistics*, eds M. Freitag, H. Kotzab & N. Megow, Springer: Cham, 2021. DOI: 10.1007/978-3-030-88662-2_8.

[26] Uimonen, S., Tukia, T., Ekström, J., Siikonen, M. & Lehtonen, M., A machine learning approach to modelling escalator demand response. *Engineering Applications of Artificial Intelligence*, **90**, 2020. DOI: 103521. 10.1016/j.engappai.2020.103521.

[27] Pallonetto, F., De Rosa, M., Milano, F. & Finn, D., Demand response algorithms for smart-grid ready residential buildings using machine learning models. *Applied Energy*, **239**, pp. 1265–1282, 2019. DOI: 10.1016/j.apenergy.2019.02.020.

[28] Singh, A.R., Ding, L., Raju, D.K., Kumar R.S. & Raghav, L.P., Demand response of grid-connected microgrid based on metaheuristic optimization algorithm. *Energy Sources, Part A: Recovery, Utilization, and Environmental Effects*, 2021. DOI: 10.1080/15567036.2021.1985654.

[29] Bui, K.-H.N., Agbehadji, I.E., Millham, R., Camacho, D. & Jung, J.J., Distributed artificial bee colony approach for connected appliances in smart home energy management system. *Expert Systems*, **37**, e12521, 2020. DOI: 10.1111/exsy.12521.

[30] Makhadmeh, S.N. et al., Multi-objective power scheduling problem in smart homes using grey wolf optimiser. *J. Ambient. Intell. Human Comput.*, **10**, pp. 3643–3667, 2019. DOI: 10.1007/s12652-018-1085-8.

[31] Ullah, I. & Hussain, S., Time-constrained nature-inspired optimization algorithms for an efficient energy management system in smart homes and buildings. Applied Sciences, **9**(4), p. 792, 2019. DOI: 10.3390/app9040792.

[32] Silva, B.N. & Han, K., Mutation operator integrated ant colony optimization based domestic appliance scheduling for lucrative demand side management. *Future Gener. Comput. Syst.*, **100**, pp. 557–568, 2019.

[33] Li, C., Jia, X., Zhou, Y. & Li, X., A microgrids energy management model based on multi-agent system using adaptive weight and chaotic search particle swarm optimization considering demand response. *Journal of Cleaner Production*, **262**, 121247, 2020. DOI: 10.1016/j.jclepro.2020.121247.

[34] Vázquez-Canteli, J., Dey, S., Henze, G. & Nagy, Z., CityLearn: Standardizing research in multi-agent reinforcement learning for demand response and urban energy management. Cornell University, 2020.

[35] Golmohamadi, H., Keypour, R., Bak-Jensen, B. & Pillai, J., A multi-agent based optimization of residential and industrial demand response aggregators. *International Journal of Electrical Power and Energy Systems*, **107**, pp. 472–485, 2019. DOI: 10.1016/j.ijepes.2018.12.020.

[36] Li, C., Liu, C., Yu, X., Deng, K., Huang, T. & Liu, L., Integrating demand response and renewable energy in wholesale market. *27th International Joint Conference on Artificial Intelligence (IJCAI)*, Stockholm, Sweden, 2018. DOI: 10.24963/ijcai.2018.53.

[37] Leo, R., Morais, A.A. & Milton, R., Advanced energy management of a micro-grid using arduino and multi-agent system. *Proceedings of ICIEES'17*, 2018. DOI: 10.1007/978-981-10-4852-4_6.

[38] Rocha, H.R.O., Honorato, I.H., Fiorotti, R., Celeste, W.C., Silvestre, L.J. & Silva, J.A.L., An artificial intelligence based scheduling algorithm for demand-side energy management in smart homes. *Appl. Energy*, **282**, 116145, 2021. DOI: 10.1016/j.apenergy.2020.116145.

[39] Grandjean, A., Adnot, J. & Binet, G., A review and an analysis of the residential electric load curve models. *Renew. Sust. Energy Rev.*, **16**, pp. 6539–6565, 2012.

[40] Zhanle, W. & Sadanand, A., Residential demand response: an overview of recent simulation and modelling applications. *IEEE 26th Canadian Conference and Computer Engineering*, Regina, SK, pp. 1–6, 2013.

[41] Gazafroudi, A.S. et al., Organization-based multi-agent structure of the smart home electricity system. *Evolutionary Computation (CEC)*, IEEE Congress, pp. 1327–1334, 2017.

[42] Pallonetto, F., De Rosa, M., Milano, F. & Finn, D., Demand response algorithms for smart-grid ready residential buildings using machine learning models. *Applied Energy*, **239**, pp. 1265–1282, 2019. DOI: 10.1016/j.apenergy.2019.02.020.

FINANCIAL TOOLS FOR BIOGAS PROJECT IMPLEMENTATION AT WASTEWATER TREATMENT PLANTS: A CASE STUDY OF THE RUSSIAN FEDERATION

ANDREY KISELEV & ELENA MAGARIL
Department of Environmental Economics, Ural Federal University, Russian Federation

ABSTRACT

Biogas projects have been recognized as one of the most efficient tools for implementing the principles of circular economy at wastewater treatment plants: negative environmental waste impact reduction along with the energy independence. These projects are widespread in developed countries, but its distribution in developing countries such as the Russian Federation leaves much to be desired. The article discusses the four financial instruments for implementation of biogas projects in the Russian Federation compared to developed countries, including the analysis of sectoral background, potential benefits, drivers and limitation. The authors concluded that barriers for the development of biogas projects are not directly related to the tool entity or its application: significant moral and physical deterioration of technological process line makes the use of these tools untimely, while the long-term payback period leads towards poor efficiency.

Keywords: circular economy, anaerobic digestion, wastewater treatment plants, financial tools, investing.

1 INTRODUCTION

Nowadays environmental situation is keenly exacerbated, due to increased anthropogenic load, exceeding the ability of the biosphere to support the process of self-regeneration. This crisis is a consequence of the practice of human consumer's behavior towards natural environment [1]. Increasing trend of global temperature growth caused by human activities via carbon and other greenhouse gases emissions gives rise to environmental instability and climate change [2].

Wastewater treatment plants (WWTPs) have been recognized as one of the significant greenhouse gases (GHG) generators, due to the complex biochemical reaction and huge consumption of energy and materials [3]. Nowadays, sustainable environmental and energy management has become significant issue in wastewater treatment sector [4]. Anaerobic digestion (AD) is well established and recognized as a robust technology to convert biomass and organic wastes to green energy and fertilizer [5] with an outstanding GHG trapping effect.

Projects based on AD techniques (so-called biogas projects) are widespread in developed countries, in contrast to developing ones. The main reason is not only the availability of modern technologies, but also the additional benefits that are realized in ensuring the energy independence of a particular country by creating a renewable energy source. It has a double profit in terms of high-energy costs in, for example, EU countries, compared to such developing countries as Russian Federation.

The key question is how to promote the circular economy principles by introducing the biogas projects in Russian Federation. It is required to consider possible financial tools of implementing the biogas projects. However, before discussing the financial tools, it is necessary to determine the main goal and specific features of these projects.

AD of sewage sludge is a treatment technique, which is designed for ensuring environmental efficiency in the management of the industrial waste. When sewage sludge is

WIT Transactions on Ecology and the Environment, Vol 255, © 2022 WIT Press
www.witpress.com, ISSN 1743-3541 (on-line)
doi:10.2495/EPM220061

disposed of in the special landfill, the organic part of it is being fermented with the release of biogas into the atmosphere, which consist of various pollutants, including carbon dioxide and methane. Both substances are greenhouse gases, but methane has a 21 times stronger effect, compared to the former one. The greenhouse effect leads towards a climate change, an increase in the frequency and scale of natural disasters, and ultimately negatively affects the habitat of humans and animals. Therefore, the capture of biogas is the most effective process to prevent the greenhouse gases from entering the atmosphere and their negative impact, which ultimately increases environmental safety in a particular area.

The biogas, obtained through anaerobic digestion process, is a valuable energy resource that can be used in various ways: to produce thermal energy in boilers, to co-generate heat and electricity in combined heat and power units, or to enrich it into biomethane for further import into the national gas transmission system.

On the other hand, biogas can be used for resource recovery through the extraction of the useful substances, such as carbon dioxide, which can be used in the chemical industry, metallurgy, food industry or elsewhere.

In terms of the sanitary and epidemiological situation, digested sewage sludge becomes stabilized: within the mesophilic mode of AD, the number of pathogens in sewage sludge is significantly reduced, while thermophilic mode of AD makes sewage sludge completely safe. The sewage sludge after AD can be used as organic fertilizer, but the application of this resource in the vast majority of areas is limited due to the presence of heavy metals [6].

The European Union adopted the Sewage Sludge Directive 86/278/EEC in 1986 and the Council Directive 91/271/EEC on urban wastewater treatment in 1991, which led towards the increase of sewage sludge disposal rates and encouraged the usage of sewage sludge in agriculture [1]. The Eurostat reports the data [7] about the agriculture use of sewage sludge as the share of total sewage sludge production. The several leaders for sewage sludge production in 2019 are presented in Table 1.

Table 1: The share of agriculture use for some EU countries in 2019 [7].

EU member	Year	Total production (1,000 tons)	Agriculture use (1,000 tons)	Share (%)
Germany	2019	1749.86	287.48	16.43
Spain*	2018	1210.40	1052.70	86.97
France**	2019	1174.00	299.00	25.47
Poland	2019	574.64	123.78	21.54
Netherlands*	2018	341.03	0.00	0.00
Austria	2019	233.56	49.70	21.28
Romania	2019	230.59	43.56	18.89
Hungary	2019	227.89	43.77	19.21
Czechia	2019	221.09	114.31	51.70
Sweden*	2018	211.60	82.30	38.89
Slovakia	2019	54.83	0.00	0.00

According to [8], in the Russian Federation, only 5%–7% of originated sewage sludge is used as a fertilizer. The share of use is negligible due to the insufficient introduction of technologies to prepare for use in agriculture. The adoption of the Federal Law of Russian Federation No. 221-FZ of June 28, 2021 "On Amendments to Certain Legislative Acts of the Russian Federation" clarified the concept of the term "agrochemicals": peat, animal and crop production waste, sludge and sewage sludge used for the production of organic and organo-mineral fertilizers were excluded from its definition [9]. It is expected that the legislative initiative will expand the opportunities for using sewage sludge as a fertilizer.

The implementation of biogas projects grants one more advantage for the company – the creation of positive company's image (or prestige). Despite the fact that this feature is quite difficult for assessment, it certainly affects many of the economic, social and manufacturing characteristics of the company.

The benefits summary for biogas project implementation is presented in Fig. 1.

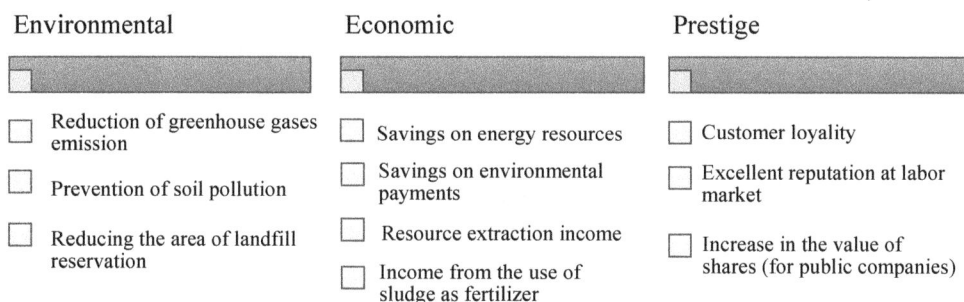

Environmental

- Reduction of greenhouse gases emission
- Prevention of soil pollution
- Reducing the area of landfill reservation

Economic

- Savings on energy resources
- Savings on environmental payments
- Resource extraction income
- Income from the use of sludge as fertilizer

Prestige

- Customer loyalty
- Excellent reputation at labor market
- Increase in the value of shares (for public companies)

Figure 1: Benefits summary for biogas project implementation.

The cost of biogas project implementation is quite significant while companies, which operate WWTP, do not have sufficient investment resources.

There are a number of studies towards the implementation of biogas technologies, their economic and environmental efficiency in the context of the transition towards sustainable development and circular economy in developed countries. However, despite the obvious advantages of biogas technologies, the feasibility of their implementation in terms of the return of investment (ROI) in Russian Federation remains uncovered. The technology of biogas yield through AD can become a niche and image building for companies as the obtained benefits do not cover (do not pay off) the invested funds.

This article discusses possible ways to finance biogas projects, with definition of the optimal tool, in the context of the feasibility of applying it to WWTPs in Russian Federation.

2 MATERIALS AND METHODS

2.1 Anaerobic digestion and combined heat and power technology

The principal technological process scheme for WWTP is presented in Fig. 2.

Wastewaters are collected from residential, commercial and industrial buildings through sewerage system and enter the receiving chamber of WWTP. Wastewaters pass through mechanical and biological treatment including solids sedimentation phase, removal of biodegradable organic matter via biochemical oxidation phase and finally go through chemical oxidation (e.g. chlorine treatment). The purified water is discharged into the water bodies, while solid, semi-solid and liquid waste is retained, concentrated and formed into sewage sludge. The originated sewage sludge is treated at mechanical dewatering workshop and then utilized at landfills [10]–[14]. The potential place for biogas projects is marked on the scheme in Fig. 2.

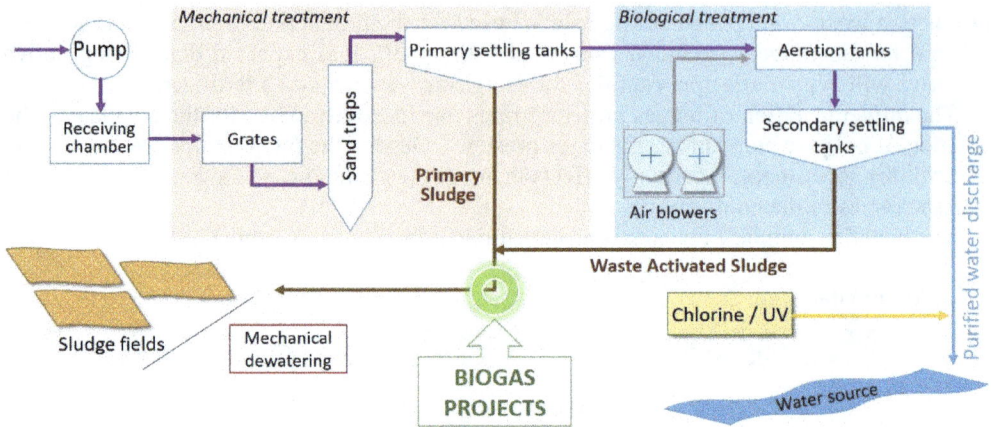

Figure 2: WWTP technological process scheme.

Anaerobic digestion is one of the most popular and widely spread environmental friendly techniques for various organic waste, including sewage sludge. In the absence of oxygen, the organic fraction of sewage sludge is degraded by microbes into biogas, which mainly consists of methane (CH_4) and carbon dioxide (CO_2) [15]–[18]. Biogas can be used directly in combined heat and power (CHP) units, burned to produce heat, or can be enriched and used in the same way as natural gas or as fuel for vehicles. The digested sewage sludge contains nitrogen, phosphorus, potassium etc. and can be applied directly or through composting as fertilizer [19], [20].

The typical technological process of anaerobic digestion is presented in Fig. 3.

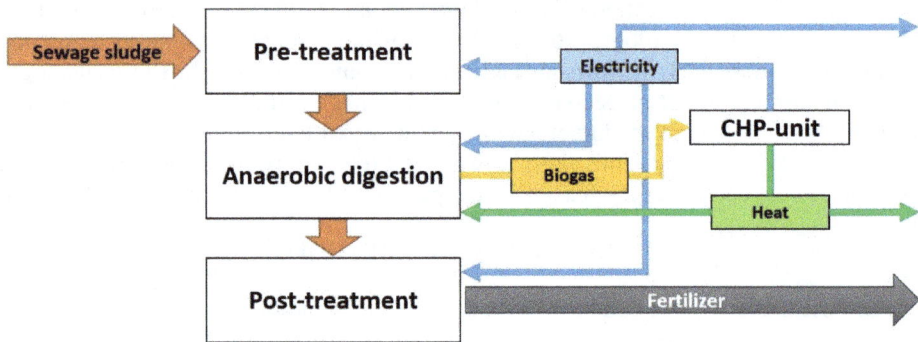

Figure 3: Typical technological process of anaerobic digestion. *(Source: Prepared by authors using data from [20].)*

The digestion process can be wet or dry, mesophilic or thermophilic, and single or multistage. A pre-treatment stage creates opportunities for improvement of biodegradability of sewage sludge through the application of different methods, including biological (e.g. bacteria or enzymes use), mechanical (e.g. ultrasonic, microwave), chemical (e.g. acids) and thermal (e.g. liquid hot water) methods. Pre-treatment improves the overall digestion process's velocity, efficiency, and sludge reduction, thereby reducing the anaerobic digester retention time and increasing the methane production rates [21], [23].

Digested sewage sludge rarely meets the discharge standards and therefore requires post-treatment. For example, digested effluent can be combined with other waste materials such as wood chip, straw or green wastes prior to composting to provide a pasteurized product – this is the best way to improve ecological safety and make valuable fertilizer. In addition, a significant amount of produced methane remains dissolved in the effluents – it can be recovered by special post-treatment techniques [22]–[25].

2.2 Case study area

Ekaterinburg is the fourth largest city in Russian Federation with the population of 1.5 billion inhabitants. It is situated on the border of Europe and Asia and located 1,667 km to the east of Moscow [26]. The centralized sewerage system of Ekaterinburg consists of two main sewerage basins: the northern and the southern. The last outpost of wastewater coming from Northern and Northeastern part of the city is the Northern wastewater treatment plant (WWTP), located in one of the industrial zones of the city. The maximum performance of the Northern WWTP is 100,000 m^3 per day.

The technological process of wastewater treatment at Northern WWTP includes primary (mechanical) and secondary (biological) treatment with ultraviolet disinfection before the discharge of treated wastewater into water bodies. Sludge treatment consists of pumping, thickening and homogenizing primary sludge and waste activated sludge, followed by anaerobic digestion in two concrete digesters, each with a volume of 5,000 m^3 (see Fig. 4), and mechanical dewatering via chamber filter press [27]. Every day almost 370 ton of the

Figure 4: Biogas project implementation at Northern WWTP.

mixture of primary sludge and waste activated sludge is loaded into methane tanks with the average hydraulic retention time of 27 days. Total biogas production for Northern WWTP in Ekaterinburg is 3.5961 t/d (is equal to 2,969.45 Nm^3/d) with consideration of biogas losses from the anaerobic digestion process of 0.0108 t/d [1].

Ekaterinburg municipal enterprise of water supply and sanitation – the operator of the Northern WWTP – is an organization carrying out regulated business activities. These types of companies provide water supply and sanitation services at tariffs set by the local authorities.

2.3 Financial tools for the project implementation

The present quality of WWTP in most of Russia's cities and towns needs improvement. According to Russia's Ministry of Natural Resources, more than 70% of the approximately 9,000 WWTPs in operation today (within centralized sewerage systems) were built 30 to 50 years ago and 80% of them should be upgraded. Furthermore, some of it cannot be upgraded, but rather must be completely rebuilt. According to the Ministry of Natural Resources, the national modernization of WWTPs will require about $20 billion, while the Ministry of Construction consider the required annual investment rate for each of the next five years for "bringing the water supply and sanitation systems into conformity with standards" at $1.5 billion [28]. In this situation, the implementation of biogas projects at WWTP is the second step in the development of the utilities. However, for consequent transition towards circular economy, the most effective financial instrument is required to be determined.

As was mentioned above, all water supply and sanitation sector is regulated, so the issue to find suitable financing is of a great importance. Four typical financial tools for biogas project implementation were determined for regulated companies, operating the WWTPs in Russian Federation. These included the following:

- tariff sources established for wastewater service users,
- state or municipal budget,
- private investment,
- bank credits.

For each tool the following steps should be fulfilled:

- to describe an organizational chart with the definition of stakeholders and their relationship in the project;
- to analyze the interest of each stakeholder in terms of the benefits from the implementation of biogas projects;
- to identify the features of financing and assess the financial efficiency of the projects;
- to consider the differences in the application of described tools in developed countries and Russian Federation.

2.4 Assessment of the financial efficiency of the biogas project

Biogas project implementation grants the initiator a multiple effect: the detailed benefits were presented in Fig. 1. However, when considering financial instruments for project implementation, the economics of the project is a crucial issue. Speaking about the use of various financial tools, Return on Investment (ROI) and Payback period (PP) indicators were applied to assess the efficiency of the projects [29].

ROI is calculated using the following equation:

$$ROI(\%) = \frac{Annual\ net\ profit}{Total\ capital\ investment} \times 100\%. \tag{1}$$

PP is calculated using the following equation:

$$PP(years) = \frac{1}{ROI}. \tag{2}$$

Annual net profit correlate with the economic benefits, mentioned above, using the following equation:

$$Annual\ net\ profit = Savings + Income - OPEX_{BGP}, \tag{3}$$

where:

- savings (\$) include savings on self-energy consumption of WWTP, savings on environmental payments;
- income (\$) includes the income from resource extraction, income from the use of digested (post-treated) sewage sludge as fertilizer and income from energy supply to the grid;
- $OPEX_{BGP}$ (\$) include operational expenditures for biogas project.

The financial equivalent of environmental and prestige factors is not taken into account under the current methodology.

Total capital investments (\$) include the required amount of investment for biogas project implementation, including design and construction works for biogas yield, CHP-unit installation, all pre- and post-treatment techniques.

3 RESULTS AND DISCUSSIONS

The authors determined the investment costs of building infrastructure for biogas yield and CHP-unit for various financial instruments for the implementation of biogas projects under consideration.

In 2018, a biogas project was put into operation at the Northern WWTP as a part of the sewage sludge treatment and utilization workshop [30]. The project was implemented according to the investment program adopted by local authorities. According to the investment program reports [31], the actual capital cost for biogas project amounted to \$4.29 million (hereinafter, all the cost values are presented in million \$ at the exchange rate for 2018).

When the biogas facility was put into operation, biogas was utilized in a heating boiler for heat energy generation. However, this disposal method has limited use only during the heating season; the rest of the time the biogas was burned in a gas-candle. The CHP-unit as an alternative method of biogas utilization was designed to improve the efficiency of biogas use. Currently, due to lack of funding, the project has not yet been implemented, but the authors will consider project data. The project included the following set of works: supply, installation and connection of a CHP-unit based on the Jenbacher JGS 312 GS-B.L generating complex with maximum electric performance of 635 kW. The project cost was \$0.69 million.

Thus, the total cost of implementing a biogas project for WWTP with a maximum inflow performance of 100,000 m^3/day was \$4.97 million.

The annual net profit within the considered case study was calculated using the following input data.

1) Self-energy consumption savings: the total annual energy generation is 4,642,800 kWh/y, which covers more than 70% of self-energy consumption of Northern WWTP.

Electricity tariffs for Russia and some EU countries, and the total amount of savings from energy generation is presented in Table 2.

The electricity tariffs in Table 2 are set for purchasing the electricity from the grid in the retail market. Nowadays in Russia, a number of regulatory documents for establishing the special "green" tariffs for electricity produced at qualified generating facilities operating as renewable energy sources, has been approved. However, by the beginning of 2021, in Russia there was the only one qualified generation unit, that use biomass [[34]]. The qualification of generating unit for renewable energy facilities according to the official requirements is too complex and possible only if self-generation exceeds the self-consumption.

Table 2: Electricity tariffs and savings in Russia and some EU countries in 2021. *(Source: Prepared by the authors, using data from [32], [33].)*

	Russia	**France**	**Italy**	**Spain**	**Germany**
Tariff ($ per kWh)	0.0676	0.2107	0.2462	0.2532	0.3480
Savings ($) m. per year	0.314	0.978	1.143	1.176	1.616
Share compared with RU	100.00%	311.81%	364.40%	374.72%	515.07%

2) Savings on environmental payments. In Russia, companies responsible for water supply and sewerage pay a fee for negative environmental impact for sewage sludge disposal: the rates depend on the hazard class of the waste [35]. In common, sewage sludge belongs to waste hazard class IV, less often – to V waste hazard class. The process of AD does not directly affect the transition from hazard class IV to V – application of post-treatment techniques is required, e.g. composting of digested sludge. However, in this case, this technology is not used. It should be mentioned, that the effect of preventing GHG from being released into the atmosphere by application of AD, does not influence a particular WWTP in terms of environmental payments.

3) Additional income. In the current case study, techniques for valuable substances extraction are not applied. These techniques, in general, have limited use in Russian Federation due to priorities: high moral and physical deterioration of the main WWT infrastructure and lack of financial resources make the use of waste recovery inappropriate.

The use of sewage sludge in Russia without any post-treatment technique may be relevant for non-agricultural applications such as quarry and landfill remediation. At present, the market for purchasing the sewage sludge for agriculture has not been established yet, so no benefits are considered.

Actual operational expenditures for biogas infrastructure at Northern WWTP are $0.17 m, for CHP-unit are $0.07 m. These costs include the personnel wages, the cost of maintenance, current repairs and spares. The total annual cost is $0.24 m. In Table 3 OPEX was presented with the approximation of the former one in some developed EU countries. The estimated share of personnel costs with accordance of average monthly wages in Russia and EC [36]–[38] was used for the approximation procedure.

The annual net profit, ROI and PP for Russia and some developed EU countries were calculated, using eqns (3), (1) and (2) respectively. The outputs are presented in Table 3.

The obtained results correlate with similar indicators from other investigations [39], [40].

The use of the various financing instruments described in Section 2.3 in relation with the investment performance indicators is presented onwards.

Table 3: $OPEX_{BGP}$, annual net profit, ROI and PP for Russia and some EU countries.

	Russia	France	Italy	Spain	Germany
$OPEX_{BGP}$ (million \$/year)	0.24	0.69	0.62	0.59	0.91
Annual net profit (million \$)	0.07	0.29	0.52	0.59	0.71
ROI (%)	1.43	5.80	10.53	11.78	14.20
PP (years)	70.13	17.24	9.50	8.49	7.04

3.1 Tariff sources established for wastewater service users

Regulatory authorities establish tariffs for wastewater services, which include transportation of wastewater via sewerage and treatment at WWTP.

The activities of Russian companies responsible for water supply and sanitation are divided into operating and investment; so, tariffs may include a production and investment component, respectively. The first step for biogas project implementation is the adjustment of water supply and sanitation scheme – a strategic planning document for municipalities. The investment costs are included in Required Gross Revenue (RGR) and the investment program passes through complex approval process. After the program is approved, tariffs are calculated and established. The annual tariff growth in Russia is limited by the level of 4%–6%.

For the case study under consideration, a forecast for tariff growth was made due to the implementation of the biogas project: the results are presented in Table 4.

Table 4: Wastewater tariff increase for biogas project implementation.

Indicator	1 year	3 year	5 year
RGR (\$ million)	4.97	1.66	0.99
Total inflow at Northern WWTP (million. cm^3)	20.72	20.72	20.72
Current (C) tariff (\$)	0.38	0.38	0.38
Additional (A) tariff (\$)	0.24	0.08	0.05
C + A tariff (\$)	0.62	0.46	0.42
Increase rate (%)	164	121	113

The calculation was made according to three options: investment costs are included in the RGR for one, three and five years.

The investment costs for biogas project implementation accounted in tariffs lead towards the annual tariff growth of 113% with the maximum RGR period of five years – it exceeds the established boundaries for tariff growth. Shorter RGR periods produce even worse results. At the same time, the implementation of other high priority investment projects is not considered. An increase in tariffs above the established values is unlikely, but possible, which will certainly cause serious dissatisfaction among consumers of services.

This financial instrument is suitable for large organizations that have significant RGR value: adding the biogas project to the investment project will have minor impact on tariff growth rate.

3.2 State or municipal budget

The EU has encouraged the development of biogas plants for energy production by various support schemes to promote transition towards circular economy. The vast majority of developed countries have legislations and energy policies related to reducing GHG and mitigation of climate change [41].

To support the development of renewables, the European Commission has established a special financing mechanism, which has been in force since September 2020. This mechanism links countries that voluntarily pay (contributing countries) with countries that agree to have new projects built on their soil (hosting countries). EU supports the financial contributors with grants, which cover either the installation of a renewable production facility (investment support) or the actual production of renewable energy (operational support) [42].

In general, in developing countries, renewable energy targets are abandoned or progress in achieving them is slow [41]. In Russia, local authorities are implementing municipal and state programs aimed at developing communal services and increasing its energy efficiency. Due to budged limitations, only few projects, predominantly critical, receive governmental support. Basically, these are projects aimed at ensuring the standard quality of wastewater treatment or preventing serious accidents at WWTPs due to high wear and tear.

Financing of biogas projects from the municipal or state budget is possible in the form of co-financing together with other sources (e.g. private capital), or interest rate compensation when using bank loans. In addition, the considered financial instrument can be used for the implementation of inter-municipal biogas projects – to solve the problem of utilization of biodegradable waste not only from WWTP, but also from other industrial enterprises, accepting animal and crop waste, municipal solid waste.

3.3 Private investment

In Russia, most WWTPs are municipal or state-owned enterprises, but recently, concession agreement procedures were approved in legislation. Government motivates investors to assist with water supply and sanitation infrastructure through rewarding them with reasonable returns [43]. Local authorities approve the parameters of the concession agreement and a list of measures to modernize the facilities. High priority has the improvement of the quality of wastewater treatment, while the significant payback period of biogas projects does not create prerequisites for the implementation of them as a private initiative to improve the energy efficiency of WWTPs.

At the same time, a number of large concessionaires have approved and are implementing a corporate social responsibility (CSR) strategy: the implementation of biogas projects affects the perception of such organizations as advanced and environmentally-friendly, both by consumers of services and by municipal and state authorities, regulatory and inspection departments.

3.4 Bank credits

Bank credit is the provision of funds by a financial institution that the borrower undertakes to repay on time with interest. Debt from private or government-owned banks is a typical and traditional financing option for biogas investment projects. The main types of these instruments for biogas projects are:

- traditional financing by loans;
- project financing (limited recourse or non-recourse), and
- green bonds [44].

The water supply and sanitation companies have to repay the loan from its activities (in Russia – from investment activities). Funding source is tariff for wastewater services. This financial instrument can be an addition for tariff financing tool reducing the annual tariff growth rate due to decreasing the value of GRG for future periods. However, long term payments (the return of loans) for biogas projects reduce the scope for other investment projects, that are not related to biogas projects.

Recently, in Russian Federation, specific banking products have appeared, aimed at the support of the development of environmental and renewable energies, including biogas projects – green credits. Large banks, particularly state-owned banks, are more likely to lend green credits. Moreover, there is a reverse relationship between green lending and the government's anti-pollution investment – as the government spends more on improving the environment, there is less need for banks to lend green credits [45].

4 CONCLUSION

Potential financial instruments for the implementation of biogas projects, including tariff sources, budged funding, private investment and bank credits were considered within the manuscript. These projects allow the consistent transition towards circular economy principles.

However, poor quality of wastewater treatment and the significant moral and physical deterioration of the utilities remain an urgent problem for WWTP managers and local authorities, responsible for communal services, in Russian Federation. These issues prevent the transition towards the next stage of WWTP modernization aimed at ensuring the environmental safety of territories and increasing the energy efficiency of facilities due to implementation of biogas projects.

Implementation of biogas projects in Russian Federation is limited by low return on investment and extremely long payback period compared to those that can be achieved in developed countries. This is the result of the low cost of energies, as well as insufficient government support for green projects aimed at developing renewable energy.

Thus, the use of the considered financial instruments in Russian Federation is insufficient and ineffective, but has made moderate progress: the state, private business and the public entities show growing interest to biogas projects every year.

ACKNOWLEDGEMENTS

The research was supported by the Russian Science Foundation grant No. 22-28-01740, https://rscf.ru/en/project/22-28-01740/.

REFERENCES

[1] Kiselev, A., Magaril, E., Magaril, R., Panepinto, D., Ravina, M. & Zanetti, M.C., Towards circular economy: Evaluation of sewage sludge biogas solutions. *Resources*, **8**, p. 91, 2019. DOI: 10.3390/resources8020091.
[2] Safari, F. & Dincer, I., Development and analysis of novel biomass-based integrated system for multigeneration with hydrogen production. *International Journal of Hydrogen Energy*, **44**(7), pp. 3511–3526, 2019. DOI: 10.1016/j.ijhydene.2018.12.101.

[3]   Kyung, D., Jung, D.-Y. & Lim, S.-R., Estimation of greenhouse gas emissions from an underground wastewater treatment plant. *Membrane Water Treatment*, **11**(3), pp. 173–177, 2020. DOI: 10.12989/mwt.2020.11.3.173.

[4]   Ranjbari, M., Esfandabadi, Z.S., Quatraro, F., Vatanparast, H., Lam, S.S., Aghbashlo, M. & Tabatabaei, M., Biomass and organic waste potentials towards implementing circular bioeconomy platforms: A systematic bibliometric analysis. *Fuel*, **318**, 123585, 2022. DOI: 10.1016/j.fuel.2022.123585.

[5]   Wu, N., Moreira, C.M., Zhang, Y., Doan, N., Yang, S., Phlips, E.J., Svoronos, S.A. & Pullammanappallil, P.C., Techno-economic analysis of biogas production from microalgae through anaerobic digestion. *Anaerobic Digestion*, ed. J. Rajesh Banu, IntechOpen, pp. 1–33, 2019. DOI: 10.5772/intechopen.86090.

[6]   Collivignarelli, M.C., Abbà, A., Miino, M.C., Caccamo, F.M., Torretta, V., Rada, E.C. & Sorlini, S., Disinfection of wastewater by UV-based treatment for reuse in a circular economy perspective. Where are we at? *International Journal of Environmental Research and Public Health*, **18**(1), pp. 1–24, 2020. DOI: 10.3390/ijerph18010077.

[7]   Eurostat, Sewage sludge production and disposal. http://appsso.eurostat.ec.europa.eu/nui/submitViewTableAction.do. Accessed on: 15 Apr. 2022.

[8]   Agrochemistry, Sewage sludge utilization. https://universityagro.ru/агрохимия/осадки-сточных-вод/. Accessed on: 15 Apr. 2022.

[9]   Federal Law of Russian Federation from 28.06.2021 № 221-FZ "On amendments to certain legislative acts of the Russian Federation". The President of Russian Federation. http://www.kremlin.ru/acts/bank/46890. Accessed on: 15 Apr. 2022.

[10]   Paśmionka, I.B. & Gospodarek, J., Assessment of the impact of selected industrial wastewater on the nitrification process in short-term tests. *International Journal of Environmental Research and Public Health*, **19**(5), p. 3014, 2022. DOI: 10.3390/ijerph19053014.

[11]   Płaza, G., Jałowiecki, Ł., Głowacka, D., Hubeny, J., Harnisz, M. & Korzeniewska, E., Insights into the microbial diversity and structure in a full-scale municipal wastewater treatment plant with particular regard to Archaea. *PLoS ONE*, **16**(4), e0250514, 2021. DOI: 10.1371/journal.pone.0250514.

[12]   Struk-Sokołowska, J., Miodoński, S., Muszyński-Huhajło, M., Janiak, K., Ofman, P., Mielcarek, A. & Rodziewicz, J., Impact of differences in speciation of organic compounds in wastewater from large WWTPs on technological parameters, economic efficiency and modelling of chemically assisted primary sedimentation process. *Journal of Environmental Chemical Engineering*, **8**(5), 104405, 2020. DOI: 10.1016/j.jece.2020.104405.

[13]   Kiselev, A., Glushankova, I., Rudakova, L., Baynkin, A., Magaril, E. & Rada, E.C., Energy and material assessment of municipal sewage sludge applications under circular economy. *International Journal of Energy Production and Management*, **5**(3), pp. 234–244, 2020.

[14]   Kiselev, A.V., Magaril, E.R. & Rada, E.C., Energy and sustainability assessment of municipal wastewater treatment under circular economy paradigm. *WIT Transactions on Ecology and the Environment*, vol. 237, WIT Press: Southampton and Boston, pp. 109–120, 2019.

[15]   Liew, C.S., Yunus, N.M., Chidi, B.S., Lam, M.K., Goh, P.S., Mohamad, M., Sin, J.C., Lam, S.M., Lim, J.W. & Lam, S.S., A review on recent disposal of hazardous sewage sludge via anaerobic digestion and novel composting. *Journal of Hazardous Materials*, **423**, Part A, 126995, 2022. DOI: 10.1016/j.jhazmat.2021.126995.

[16] Farida, H., Chang, Y.L., Hirotsugu, K., Abdul, A.H., Yoichi, A., Takeshi, Y. & Hiroyuki, D., Treatment of sewage sludge using anaerobic digestion in Malaysia: Current state and challenges. *Frontiers in Energy Research*, 7, 2019. DOI: 10.3389/fenrg.2019.00019.

[17] Akbay, H.E.G., Deniz, F., Mazmanci, M.A., Deepanraj, B. & Dizge, N., Investigation of anaerobic degradability and biogas production of the starch and industrial sewage mixtures. *Sustainable Energy Technologies and Assessments*, **52**, Part A, 2022. DOI: 10.1016/j.seta.2022.102054.

[18] Palermito, F., Magaril, E., Conti, F., Kiselev, A. & Rada, E.C., Circular economy concepts applied to waste anaerobic digestion plants. *WIT Transactions on Ecology and the Environment*, vol. 254, WIT Press: Southampton and Boston, pp. 57–68, 2021.

[19] Hidaka, T., Nakamura, M., Oritate, F. & Nishimura, F., Utilization of high solid waste activated sludge from small facilities by anaerobic digestion and application as fertilizer. *Water Science and Technology*, **80**(12), pp. 2320–2327, 2019. DOI: 10.2166/wst.2020.050.

[20] Department for Environment, Food and Rural Affairs, UK, Anaerobic digestion strategy and action plan: A commitment to increasing energy from waste through anaerobic digestion. https://assets.publishing.service.gov.uk/government/uploads/system/uploads/attachment_data/file/69400/anaerobic-digestion-strat-action-plan.pdf. Accessed on: 15 Apr. 2022.

[21] Park, S., Yoon, Y.-M., Han, S.K., Kim, D. & Kim, H., Effect of hydrothermal pre-treatment (HTP) on poultry slaughterhouse waste (PSW) sludge for the enhancement of the solubilization, physical properties, and biogas production through anaerobic digestion. *Waste Management*, **64**, pp. 327–332, 2017. DOI: 10.1016/j.wasman.2017.03.004.

[22] Ekstrand, E.M. et al., Identifying targets for increased biogas production through chemical and organic matter characterization of digestate from full-scale biogas plants: What remains and why? *Biotechnology for Biofuels and Bioproducts*, **15**, p. 16, 2022. DOI: 10.1186/s13068-022-02103-3.

[23] Fernández-Polanco, D., Aagesen, E., Fdz-Polanco, M. & Pérez-Elvira, S.I., Comparative analysis of the thermal hydrolysis integration within WWTPs as a pre-, inter- or post-treatment for anaerobic digestion of sludge. *Energy*, **223**, 120041, 2021. DOI: 10.1016/j.energy.2021.120041.

[24] Rongwong, W. et al., Membrane-based technologies for post-treatment of anaerobic effluents. *Clean Water*, **1**, p. 21, 2018. DOI: 10.1038/s41545-018-0021-y.

[25] Fan, Y.V., Klemeš, J.J. & Lee, C.T., Pre- and post-treatment assessment for the anaerobic digestion of lignocellulosic waste: p-graph. *Chemical Engineering Transactions*, **63**, pp. 1–6, 2018. DOI: 10.3303/CET1863001.

[26] Britannica, Yekaterinburg. https://www.britannica.com/place/Yekaterinburg. Accessed on: 15 Apr. 2022.

[27] Ekaterinburg Municipal Enterprise of Water Supply and Sanitation, Idem na sever: About Northern WWTP. https://www.водоканалекб.рф/путь-воды/идем-на-север. Accessed on: 15 Apr. 2022. (In Russian.)

[28] Wisconsin Economic Development Corporation (WEDC), Russia prepares to modernize water treatment plants, February 2019. https://wedc.org/export/market-intelligence/posts/russia-prepares-to-modernize-water-treatment-plants/. Accessed on: 15 Apr. 2022.

[29] Buller, L.S., Sganzerla, W.G., Berni, M.D., Brignoli, S.C. & Forster-Carneiro, T., Design and techno-economic analysis of a hybrid system for energy supply in a wastewater treatment plant: A decentralized energy strategy. *Journal of Environmental Management*, **305**, 114389, 2022. DOI: 10.1016/j.jenvman.2021.114389.

[30] A complex of digesters for biogas production was put into operation at the northern aeration station of Yekaterinburg. https://watermagazine.ru/novosti/proekty/22003-na-severnoj-aeratsionnoj-stantsii-g-ekaterinburga-vveden-v-ekspluatatsiyu-kompleks-metantenkov-po-vyrabotke-biogaza.html. Accessed on: 15 Apr. 2022.

[31] Reports on the implementation of the investment program of MUE "Vodokanal" in the framework of information disclosure standards. https://www.водоканалекб.рф/раскрытие-информации/раскрытие-информации/водоотведение.

[32] Eurostat, Electricity price statistics. https://ec.europa.eu/eurostat/statistics-explained/index.php?title=Electricity_price_statistics. Accessed on: 15 Apr. 2022.

[33] Electricity tariffs in Sverdlovsk region. https://ekb.esplus.ru/tariffs/ekb/fiz/elektroenergiya/tarify/. Accessed on: 15 Apr. 2022.

[34] Association NP Market Council, Review of cents (tariffs) for electricity produced at qualified generating facilities operating on the basis of renewable energy sources for 2021. https://www.np-sr.ru/sites/default/files/obzor_vie_2021_.doc. Accessed on: 15 Apr. 2022.

[35] Decree of the Government of the Russian Federation of September 13, 2016 No. 93 "On the rates of payment for the negative impact on the environment and additional coefficients". https://docs.cntd.ru/document/420375216. Accessed on: 15 Apr. 2022.

[36] Average and minimum salaries in Europe: Wages levels for the EU countries and some other countries of the world, salary tables for 2022. https://ru-geld.de/en/salary/europe.html. Accessed on: 15 Apr. 2022.

[37] Eurostat, Wages and labour costs. https://ec.europa.eu/eurostat/statistics-explained/index.php?title=Wages_and_labour_costs. Accessed on: 15 Apr. 2022.

[38] Ministry of Economics and Spatial Development, The information about average salary in Sverdlovsk region in January–August 2021. http://economy.midural.ru/content/informaciya-o-srednemesyachnoy-zarabotnoy-plate-v-sverdlovskoy-oblasti-v-yanvare-avguste-4. Accessed on: 15 Apr. 2022.

[39] Niekurzak, M., The potential of using renewable energy sources in Poland taking into account the economic and ecological conditions. *Energies*, **14**, p. 7525, 2021. DOI: 10.3390/en14227525.

[40] Nami, H., Anvari-Moghaddam, A. & Arabkoohsar, A., Thermodynamic, economic, and environmental analyses of a waste-fired trigeneration plant. *Energies*, **13**, p. 2476, 2020. DOI: 10.3390/en13102476.

[41] Chrispim, M.C., Scholz, M. & Nolasco, M.A., Biogas recovery for sustainable cities: A critical review of enhancement techniques and key local conditions for implementation. *Sustainable Cities and Society*, **72**, 103033, 2021. DOI: 10.1016/j.scs.2021.103033.

[42] EU renewable energy financing mechanism. https://energy.ec.europa.eu/topics/renewable-energy/financing/eu-renewable-energy-financing-mechanism_en. Accessed on: 15 Apr. 2022.

[43] Yu, S., Dai, T., Yu, Y. & Zhang, J., Investment decision model of wastewater treatment public–private partnership projects based on value for money. *Water and Environment Journal*, **35**(1), pp. 322–334, 2021. DOI: 10.1111/wej.12629.

[44] Renewable Gas Trade Centre in Europe (REGATRACE), D6.2 | Guidebook on securing financing for biomethane investments. https://www.regatrace.eu/wp-content/uploads/2020/12/REGATRACE-D6.2.pdf. Accessed on: 15 Apr. 2022.

[45] Yin, W., Zhu, Z., Kirkulak-Uludag, B. & Zhu, Y., The determinants of green credit and its impact on the performance of Chinese banks. *Journal of Cleaner Production*, **286**, 124991, 2021. DOI: 10.1016/j.jclepro.2020.124991.

SECTION 2
ENERGY AND
SUSTAINABILITY

ANALYSIS OF THE WATER–ENERGY–FOOD NEXUS AND ITS CONTRIBUTION TO ENERGY DEVELOPMENT

GRICELDA HERRERA-FRANCO[1,2], HARRY ALBERTO BOLLMANN[3],
JANAINA CAMILE PASQUAL LOFHAGEN[3] & CARLOS MORA-FRANK[4]
[1]Programa de Pós-graudação em Gestão Urbana, Pontifícia Universidade Católica do Paraná (PUCPR), Brazil
[2]Facultad de Ciencias de la Ingeniería, Universidad Estatal Península de Santa Elena (UPSE), Ecuador
[3]Pontifícia Universidade Católica do Paraná (PUCPR), Brazil
[4]Centro de Investigación y Proyectos Aplicados a las Ciencias de la Tierra (CIPAT),
Escuela Superior Politécnica del Litoral, ESPOL, Ecuador

ABSTRACT
The global energy system is moving towards a sustainable future with new development strategies that reduce the carbon footprint, such as the water–energy–food (WEF) nexus. Several countries have implemented this link to provide energy and food security while maintaining the relationship between socio-economic progress and environmental protection. The WEF nexus with energy development generates new interest in innovation, and it is important to explore the growth of this academic field. The work aims to analyse the scientific development of the WEF nexus during energy intervention processes, through bibliometric review models, for the knowledge of strategies in a bioenergy framework. The methodology consists of: (i) information compilation (Scopus and Web of Science) and software selection; (ii) information review on scientific production, author keywords and countries; and (iii) focus group analysis in a framework of energy development. The results show scientific interest from 2007, with exponential growth from 2016. The literature presents the interest of implementing the WEF nexus in energy processes to reduce environmental pollution, like ethanol in gasoline, biorefineries, sustainable agriculture, hydropower, and renewable energies (solar, wind). This scientific approach is dominated by the USA, China and the United Kingdom in environmental science, energy and engineering areas, accounting for 60% of the production. The study shows that the WEF nexus approach to energy developments creates new prospects for decision-making in socio-economic, political, and environmental progress.
Keywords: WEF nexus, bioenergy, sustainability, bibliometric analysis.

1 INTRODUCTION
Energy sources play an essential role in the socio-economic progress of developed and developing countries. Currently, non-renewable energies such as oil and natural gas are considered a resource that sustains modern society and today's economy (fuel consumption and industrial processes) [1], [2]. These energies cover 33.1% of global consumption, generating economic and social benefits [3]. However, estimates indicate a limited reserve for supply from these resources [4].

Although the oil industry integrates sustainability approaches into its activities, there are problems with energy and environmental issues, such as energy demand growth and climate change [5], [6]. Therefore, global organizations have developed renewable energy sources under the sustainability framework for socio-economic development, strengthening environmental protection, and substituting fossil fuels.

Renewable energies are sustainable methods that use the natural resources of an environment, like the dynamics of water (hydropower), the sun heat (geothermal energy), the wind power (wind energy), and the use of biological sources (bioenergy) [7], [8]. In this case, bioenergy is considered a renewable resource that comes from materials of biological origin, like plants or plant derivatives. Bioenergy is the second-largest commercial source of renewable energy after hydropower and is characterized by high efficiency, cleanliness, and convenience in the commercial sector [9]. For example, biomass is considered a distinctive and promising type of green energy that has a single-use and can be converted into a biofuel energy source [10], [11]. This advantage allows biofuels to receive much attention over the last two decades from scientists worldwide, with changes in their pace of development, focus, and international collaboration. In addition, bioenergy has a crucial strategic contribution to new energy and economic challenges, like reducing fossil energy use and socio-economic development [12]. Bioenergy production has two phases for the generation of products (e.g., bioethanol, biodiesel, biogas, butane, dimethylether, methanol, hydrogen, vegetable oil): (i) preparation, related to cultivation, harvesting, collection, and (ii) processing, referred to esterification, fermentation, ficher, and gasification [13].

Studies highlight the importance of strengthening a strategic system with renewable energies, mainly in areas with higher socioeconomic and industrial demand and places with abundant biodiversity [14]. In recent years, the water, energy, and food (WEF) nexus have been promoted through international meetings and convenings as a global research agenda and an emerging sustainable development paradigm [15]. The importance of the WEF nexus arises from the scarcity problems of natural resources like water, energy, food, and resources related to human needs [16], [17]. Therefore, the water–energy nexus relationship was established from the importance of water for energy generation in the water supply processes and wastewater treatment, while the food–water nexus relates to crop irrigation for agricultural activities [18], [19].

WEF nexus strengthens the concept of sustainability in resources and management. However, there are limited policy responses and resource constraints in the processes of industrialisation, trade and consumption, with the implementation of the "planning" and "decision-making" frameworks being strategic [20], [21]. Moreover, the close relationship between the WEF nexus and bioenergy security creates problems in the nexus, directly related to food and water security. For example, biofuels used for transport in a country with a high energy demand create the need to occupy new territories to increase the production of biological material [22]. The importance of the WEF nexus in policies in various sectors (e.g. agriculture, water, energy, climate) determines the trade-off outcomes for society and the environment [23]. According to Mirzabaev et al. [24], studying the bioenergy economy in the WEF nexus framework helps a clear picture of positive synergy opportunities and potential constraints.

The WEF nexus information with bioenergy shows the need to incorporate a general analysis of the findings and scientific contribution that countries around the world contribute to the evolution of this study field. Therefore, bibliometric analysis is a type of research that makes it possible to visualise or represent trends related to scientific development [25]. In addition, bibliometrics has contributed to various fields of research, like medicine [26], industry [27], engineering [28], and the relationship between oil and the environment [29]. These analyses use systematic tools or techniques that make it easier to obtain bibliographic information by searching databases like Scopus and Web of Science (WoS) [30]. Scopus has extensive coverage of scientific disciplines, ease of access and visualisation [31], and WoS provides older research with approximately 1,300 prestigious journal articles [32], and 256 disciplines [33].

This leads us to establish some concerns: How to visualise the intellectual structure and the various fields of WEF nexus in the context of bioenergy? Which scientific approaches and countries are most interested in strengthening the WEF nexus concept in energy development?

The work aims to analyze the scientific development of the WEF nexus in energy intervention processes worldwide through bibliometric review models in the Scopus and

WoS databases to know energy development strategies with expert opinions in the scientific area.

2 METHODOLOGY

This research was developed in three phases (Table 1): (i) information compilation in Scopus and WoS databases and use of software (e.g., VOSviewer and Bibliometrix); (ii) review and analysis of the information, generating graphs of scientific production, author keywords, and countries; and (iii) focus group analysis in a framework of energy development.

Table 1: Methodological outline of the study.

Phase I	**1. Information compilation** - *Database:* Scopus and WoS. - *Research topic:* "water-energy-food" OR "water-food-energy" OR "food-energy-water" OR "WEF nexus" OR "WFE nexus" OR "FEW nexus" AND "biofuel*" OR "bioenergy*" OR "biomass". **2. Softwares used** - *VOSviewer:* integral bibliometric mapping. - *Bibliometrix:* data fusion, data cleansing and reference data.
Phase II	**3. Data analysis** - Scientific production of study field. - Author keywords network. - Keyword Frequency. - Scientific production of countries and authors.
Phase III	**4. Focus group analysis** - Expert opinion in the field, based on the framework of energy development.

2.1 Information gathering and software selection

This work is based on the systematic search and structuring of information related to the WEF nexus in the bioenergy field. The collection considers publications of articles, conference papers, books, book chapters and others included in databases (i.e., Scopus and WoS). The search criteria consider the WEF nexus and similar criteria, intercepting bioenergy with biofuel and biomass criteria. For example, the topic search is "water–energy–food" OR "water–food–energy" OR "food–energy–water" OR "WEF nexus" OR "WFE nexus" OR "FEW nexus" AND "biofuel*" OR "bioenergy*" OR "biomass", searching on titles, abstracts, and keywords. In addition, a data filtering mechanism is included, like inclusion terms that consider all types of documents, sources, languages and subject areas, and exclusion terms that involve removing production from the year 2022 (current year). These conditions grouped 139 Scopus publications and 145 WoS publications.

Once the search criteria and conditions are defined, the information is downloaded in Comma-Separated Values (CSV), BibTex, Plan Text File and Excel format. These formats presented inconsistent records, which necessitated a detailed cleaning of the data, eliminating duplicate and missing files (author name, title) – obtaining 138 documents from Scopus and 145 from WoS.

The study used bibliometric management software for data processing, like Visualisation of Similarity Viewer (VOSviewer) and Bibliometrix. VOSviewer is a freely available computer programme that allows the construction and visualisation of connection maps, like author keywords, and countries [34]. This software generated the author keywords map through data extraction-processing and cleaning the information with thesaurus [35], [36]. On the other hand, Bibliometrix is an RStudio package that allows data processing through specific tools [37]. The present study used RStudio version 4.1.2 and the functions readfile/convert2df. This programme made it possible to merge Scopus and WoS data using specific coding and formats (e.g., BibTex and Plain Text File). The combination of data showed 25 unique Scopus documents, 32 unique WoS documents and 113 in both databases, resulting in a total merger of 170 publications that were used in the analyses.

2.2 Analysis of data

This section generates the analysis of the obtained information containing the intellectual approach and its links, considering the scientific production, author keywords network, keywords frequency, and countries–authors relationship. In a first analysis, the scientific output shows the quantitative distribution of publications generated over time. Subsequently, the graph of author keywords allows us to visualise the lines of research in this study field, while the frequency of keywords shows the initial and final evolution. Finally, the last analysis presents the nations and authors who developed this field of research considerably.

2.3 Focus group analysis

The focus group analysis shares scientific opinion content from expert authors in the field, providing a scientific content of quantitative context for the social and academic community [38]. In general, this analysis values the interaction of participants, presenting tactics or strategies to solve problems and transform realities [39].

3 RESULTS

3.1 Analysis of data

3.1.1 Scientific production of the study field
Publications related to the WEF nexus in the bioenergy field is a topic of recent integration in the academic world. Fig. 1 shows that the production started in 2007 with only one publication, maintaining a linear growth until 2014. According to Price's Law [40], the trend becomes exponential from that year onwards. This increase indicates the interest of scientists in integrating the WEF nexus during energy development (in this case, bioenergy), as the topic of friendly energies becomes an important sustainable mechanism in energy generation and multiple water use activities.

3.1.2 Countries–authors relationship
The analysis shows the intersection between the authors and countries, allowing us to know the intellectual contribution. The author keywords are also added to determine their lines of research (Fig. 2). In general, 48 countries and 693 authors have conducted WEF nexus studies related to the bioenergy field, led by the USA, Malaysia, Brazil, and China. USA presents a contribution of 135 publications with six relevant authors in 12 important scientific lines, obtaining prominent participation with the WEF nexus and bioenergy. Although Malaysia is ranked second with 73 contributions, it shows lower participation than other nations, as it has a contribution in six keywords. However, its five principal authors have an outstanding

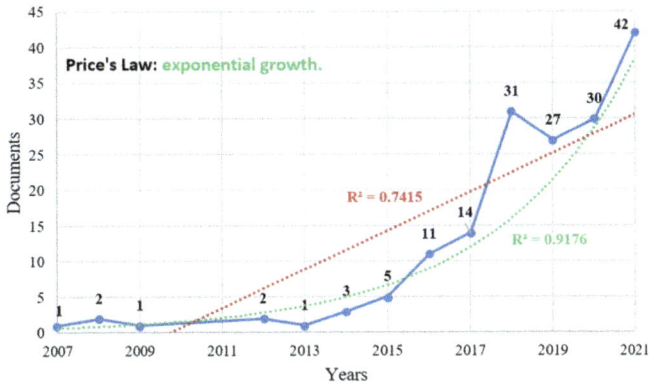

Figure 1: WEF nexus scientific output in the bioenergy field.

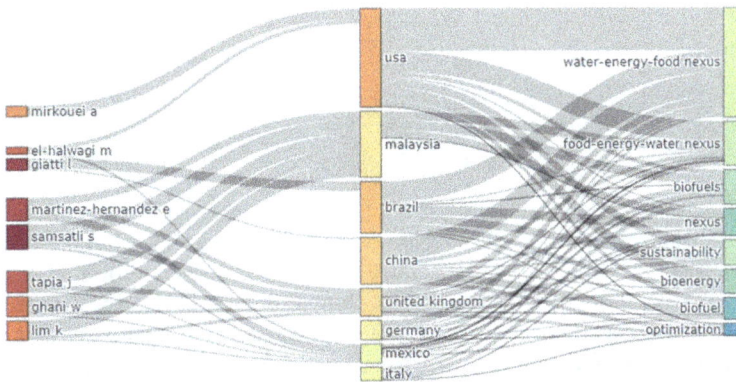

Figure 2: Author–country relationship and lines of research in scientific production.

scientific performance. Brazil and China reflect a close production (61 and 57 papers, respectively) with essential authors in this field of research, such as L. Zhang and L. Giatti. Other countries also contribute to scientific development, like the United Kingdom, Germany, Mexico, and Italy.

3.1.3 Author keywords network

The VOSviewer programme allowed the construction and visualisation of this map by processing 518 keywords. In addition, the cleaning mechanism (thesaurus) and the occurrence conditions (at least four times) presented a total of 26 keywords (nodes), constituting six clusters on the map (Fig. 3).

Cluster 1: "Sustainability and biomass" (red colour), has the highest number of nodes (seven) with 78 occurrences. This cluster shows the outstanding involvement of the food–energy–water (FEW) nexus in biomass relations, sustainability, and its optimisation in these concepts. The information shows the importance of sustainability in the nexus with the framework of socio-economic and environmental activities [41]. In addition, life cycle assessment (LCA) strengthens environmental impact analyses, considering the FEW nexus in energy crop production [42].

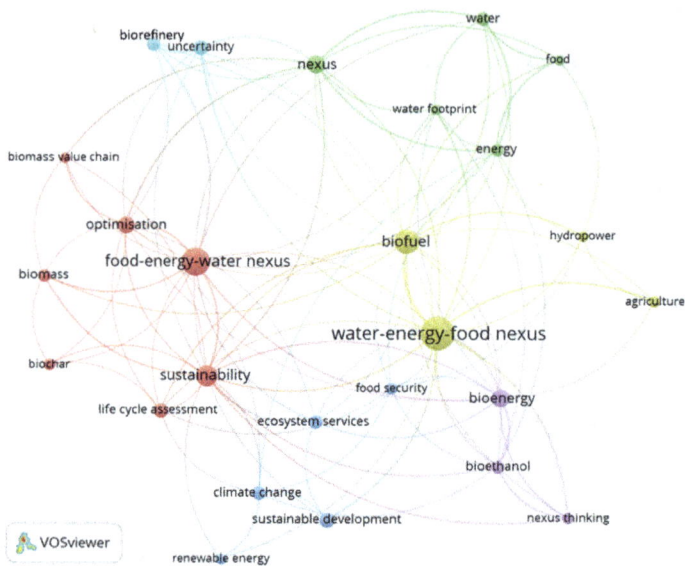

Figure 3: Author keywords connection map.

Cluster 2: "Nexus and water footprint" (green colour), has five nodes totalling 32 occurrences. The cluster shows a distribution of the WEF nexus linked to the water footprint and slightly to food security for multiple water use and energy generation analyses [43].

Cluster 3: "Security and climate change" (blue colour) indicates the connection of five nodes and 31 occurrences in total. In this cluster, food security has a higher magnitude and a solid link to the various types of bioenergy (biofuel and bioethanol), as climate change issues indicate the importance of integrating new sustainability scopes (e.g., WEF nexus and renewable energies).

Cluster 4: "Biofuels" (yellow colour), is representing a set of four nodes (69 occurrences). This cluster indicates the critical contribution of the WEF nexus in using biofuels, as despite containing bio-products (biomass), it generates environmental impacts on food sources and economic capital [44]. In addition, studies highlight that biomass is a viable option for a climate-friendly and energy-efficient economy [45].

Cluster 5: "Bioenergy and nexus integration" (purple colour), has three nodes totalling 22 occurrences. The cluster terms show studies that analyse energy production and its association with bio-products to understand the integration of the WEF nexus in the development of this field [46].

Cluster 6: "Control and bioprocesses" (light blue), has the lowest number of nodes (two) with a total of 15 occurrences. The cluster contains studies indicating biorefineries' importance in obtaining biofuels to cover the bioenergy demand. However, other studies confirm the environmental problems of biorefineries in hydrographic systems [47].

3.1.4 Keywords frequency
This section shows the keywords that appear in at least three studies, placing the node in the year of highest frequency (Fig. 4) [48]. The data shows that the hydropower power generation industries dominated this field in relationship with other methods, with the highest number

of hydropower studies in 2013. In later years, energy production shifted with great interest towards renewable energies and biofuels, as the issue of climate change caused ecosystemic consequences like a lack of water resources for agricultural irrigation and multiple water use. The nexus concepts, sustainability and LCA, are also beginning to be involved in coherence with bioenergy growth to strengthen environmental control during the processes and use of these energies. This relationship arises from the contamination of ecosystems caused by biorefineries and the emission of toxic gases generated by biofuels due to the lack of control during blending with bio-products (low percentage). Finally, the information reflects current relevant topics (2020–2021), like food waste, integrated assessment modelling, nexus thinking, and bioethanol.

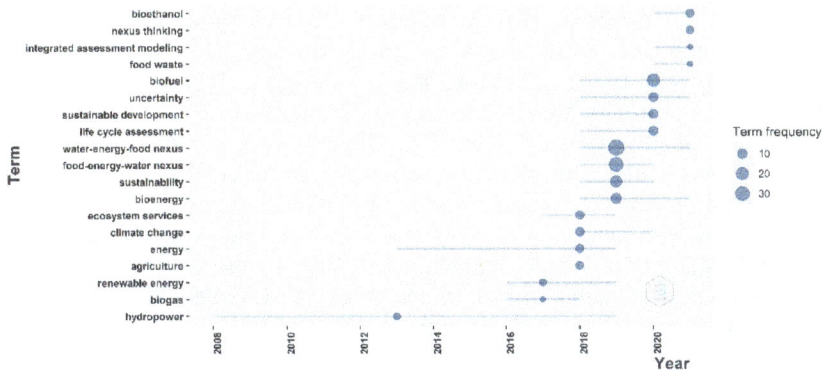

Figure 4: Evolution of the main keywords over time.

3.2 Analysis focus group

Although this scientific field is of recent interest, the information gathered shows the importance of including the WEF nexus in current energy development topics. This relationship strengthens the concept of sustainability in socio-economic activities and environmental protection in a framework of water and food demand due to global population growth. However, this criterion does not show significant results in nations worldwide, mainly developing countries. Therefore, experts consider it essential to implement energy and socio-economic development strategies, like (i) the use of renewable energies available in the natural environment of a locality (solar, wind and water energy), (ii) rescue of water ancestral knowledge, (iii) wastewater use with the implementation of tertiary treatment systems in a context of monitoring and control for water reuse, (iv) natural areas of protection mangrove and bird, known as green dots, (v) enhancing waste management like reuse, composting, biogas, and mechanical treatment.

4 DISCUSSION

The Scopus and WoS databases allowed the information download through a search strategy for data fusion and generation of analyses. However, the authors consider it necessary to clean up the records, as there are inconsistencies in the download, like duplicates and records without author or title information. This cleaning process presents quality and reliable results to understand better the scientific output that addresses WEF nexus studies in energy development. Some publications use Scopus and WoS tools for filtering and subsequent data cleaning, allowing the development of a complete content review [49].

This bibliometric study indicates the scientific interest in integrating the WEF nexus with bioenergy in the last three years (2019–2021). According to Han et al. [50], the WEF nexus faces challenges in developing the concept in science and policy areas of intervention for economic development. Therefore, our results reflected the essentials of the WEF nexus in energy management and multiple water use related to the water footprint and climate change. Hamidov and Helming [51] indicate a strategic conceptual synthesis of the WEF nexus and that it annually strengthens its operation on sustainability issues.

On the other hand, it is essential to mention the strategies found in this study related to the WEF nexus in the development of bioenergy, based on the keywords frequency and author keywords network: (i) the promotion of including life cycle assessment in environmental impact analyses and the WEF nexus in socio-economic development, generates a fundamental relationship in the promotion and monitoring of bioenergy; (ii) food security must have a direct link to the WEF nexus concept, as water use is currently not sustainably managed; (iii) sustainable projects related to climate change should be developed and implemented in developing countries, integrating the direct collaboration of international experts; (iv) the implementation and promotion of biofuels is an affordable way forward for emerging countries; (v) countries with a demand for water, energy and food develop a legal link with global organisations for strengthening the WEF nexus in energy development (e.g. bioenergy); (vi) the sustainable development goals (SDGs) should be integrated more frequently into studies in this field of research.

5 CONCLUSIONS

The scientific production analysis of the WEF nexus with bioenergies shows a recent interest of 14 years (2007–2021) with 170 publications when merging information from Scopus and WoS, and an exponential growth reflected in the last four years (76.4% between 2018–2021). In addition, we highlight the collaboration of 48 countries in various studies, led by the USA and Malaysia. There are also a general 518 keywords, interpreting the central nodes of each cluster that focus on topics like sustainability, biomass, nexus connection and integration, security, climate change, bioprocesses, and bioenergy.

The WEF nexus is an important concept that needs to be integrated more frequently into the energy development of the world's countries, maintaining the link to sustainability. This relationship becomes a fundamental topic in decision-making processes for sustaining the energy–water nexus demand in agricultural systems, municipal and economic environments. On the other hand, the strategies implemented would reinforce this context within an environmental protection and social development framework, highlighting that the themes of current interest are food waste, integrated assessment modelling, nexus thinking, and bioethanol. In addition, the promotion of new sustainable methods such as LCA and the SDGs are crucial in linking them to the WEF nexus and bioenergies.

ACKNOWLEDGEMENTS

This work was carried out with the collaboration of the "Peninsula Santa Elena Geopark Project" with code no. 91870000.0000.381017, and "Factores Geoambientales de los pozos petroleros y su incidencia en el desarrollo territorial en los cantones Salinas y La Libertad de la provincia de Santa Elena", with code no: 91870000.0000.385428, of the UPSE University project (Universidad Estatal Península de Santa Elena). Furthermore, support for the academic research project "Registry of geological and mining heritage and its impact on the defense and preservation of geodiversity in Ecuador" by ESPOL University is gratefully acknowledged, CIPAT-01-2018.

REFERENCES

[1]   Vassiliou, M. (eds), *Historical Dictionary of the Petroleum Industry*, Rowman & Littlefield: Lanham, USA, p. 593, 2018.

[2]   Fanchi, J.R. & Fanchi, C.J. (eds), *Energy in the 21st Century,* World Scientific Publishing Company: Danvers, USA, p. 469, 2016.

[3]   Looney, B., BP statistical review of world energy. www.bp.com/en/global/corporate/ energy-economics/statistical-review-of-world-energy.html. Accessed on: 29 Mar. 2021.

[4]   Öz, S. & Alyürük, M., Energy sector overview and future prediction for Turkey. *Journal of Industrial Policy and Technology Management*, 3(1), pp. 59–69, 2020.

[5]   Asmoro, T.H., Prabowo, H.E. & Prayascitra, A., Lessons learned from adopting sustainability aspects for gas development project. *SPE/IATMI Asia Pacific Oil and Gas Conference and Exhibition*, 2020.

[6]   Herrera-Franco, G., Escandón-Panchana, P., Erazo, K., Mora-Frank, C. & Berrezueta, E., Geoenvironmental analysis of oil extraction activities in urban and rural zones of Santa Elena Province, Ecuador. *International Journal of Energy Production and Management*, 6(3), pp. 211–228, 2021.

[7]   Afgan, N.H., Al Gobaisi, D., Carvalho, M.G. & Cumo, M., Sustainable energy development. *Renewable and Sustainable Energy Reviews*, 2(3), pp. 235–286, 1998.

[8]   Herrera-Franco, G., Carrión-Mero, P., Aguilar-Aguilar, M., Morante-Carballo, F., Jaya-Montalvo, M. & Morillo-Balsera, M.C., Groundwater resilience assessment in a communal coastal aquifer system. The case of Manglaralto in Santa Elena, Ecuador. *Sustainability*, 12(19), p. 8290, 2020.

[9]   IES Bioenergy, *The Role of Bioenergy in Greenhouse Gas Mitigation*, IES Bioenergy, 1998.

[10]   Long, H., Li, X. & Wang, H., Biomass resources and their bioenergy potential estimation: A review. *Renewable and Sustainable Energy Reviews*, 26, pp. 344–352, 2013.

[11]   Ozturk, M., Saba, N., Altay, V., Iqbal, R., Rehman Hakeem, K., Jawaid, M. & Hanum Ibrahim, F., Biomass and bioenergy: An overview of the development potential in Turkey and Malaysia. *Renewable and Sustainable Energy Reviews*, 79, pp. 1285–1302, 2017.

[12]   Azadi, P., Malina, R., Barrett, S.R. & Kraft, M., The evolution of the biofuel science. *Renewable and Sustainable Energy Reviews*, 76, pp. 1479–1484, 2017.

[13]   Jeswani, H.K., Chilvers, A. & Azapagic, A., Environmental sustainability of biofuels: A review. *Proceedings of the Royal Society A*, 476(2243), p. 37, 2020.

[14]   Rehman Hakeem, K., Jawaid, M. & Rashid, U. (eds), *Biomass and Bioenergy: Applications*, Springer: Selangor, Malaysia, p. 405, 2014.

[15]   Leck, H., Conway, D., Bradshaw, M. & Rees, J., Tracing the water–energy–food nexus: Description, theory and practice. *Geography Compass*, 9(8), pp. 445–460, 2015.

[16]   Dupar, M. & Oates, N., Getting to grips with the water–energy–food 'nexus'. Climate and Development Knowledge Network (London), 2012. https://odi.org/en/insights/ getting-to-grips-with-the-water-energy-food-nexus/. Accessed on: 24 Mar. 2022.

[17]   Hoff, H., Understanding the nexus. *Bonn2011 Nexus conference: The Water, Energy and Food Security Nexus*, Stockholm Environment Institute (SEI), 2011.

[18] Kirchem, D., Lynch, M.Á., Bertsch, V. & Casey, E., Modelling demand response with process models and energy systems models: Potential applications for wastewater treatment within the energy–water nexus. *Applied Energy*, **260**, 114321, 2020.

[19] Heard, B.R., Miller, S.A., Liang, S. & Xu, M., Emerging challenges and opportunities for the food–energy–water nexus in urban systems. *Current Opinion in Chemical Engineering*, **17**, pp. 48–53, 2017.

[20] Pittock, J., Hussey, K. & McGlennon, S., Australian climate, energy and water policies: Conflicts and synergies. *Australian Geographer*, **44**(1), pp. 3–22, 2013.

[21] Pasqual, J.C., Lardizabal, C.C., Herrera, G., Bollmann, H.A. & Nunes, E.O., Water–energy–food nexus: Comparative scenarios and public policy perspectives from some Latin American countries regarding biogas from agriculture and livestock. *Journal of Agricultural Science and Technology A*, **5**(6), pp. 408–427, 2015.

[22] Asian Development Bank (ADB) (eds), *Thinking About Water Differently: Managing the Water–Food–Energy Nexus*, Asian Development Bank: Philippines, p. 47, 2013.

[23] Ringler, C., Bhaduri, A. & Lawford, R., The nexus across water, energy, land and food (WELF): Potential for improved resource use efficiency? *Current Opinion in Environmental Sustainability*, **5**(6), pp. 617–624, 2013.

[24] Mirzabaev, A., Guta, D., Goedecke, J., Gaur, V., Börner, J., Virchow, D., Denich, M. & von Braun, J., Bioenergy, food security and poverty reduction: Trade-offs and synergies along the water–energy–food security nexus. *Water International*, **40**(5–6), pp. 772–790, 2015.

[25] Zupic, I. & Čater, T., Bibliometric methods in management and organization. *Organizational Research Methods*, **18**(3), pp. 429–472, 2015.

[26] Tran, B. et al., Global evolution of research in artificial intelligence in health and medicine: A bibliometric study. *Journal of Clinical Medicine*, **8**(3), p. 360, 2019.

[27] Mei, Y., Ma, T. & Su, R., How marketized is China's natural gas industry? A bibliometric analysis. *Journal of Cleaner Production*, **306**, 127289, 2021.

[28] Cancino, C., Merigó, J.M., Coronado, F., Dessouky, Y. & Dessouky, M., Forty years of *Computers and Industrial Engineering*: A bibliometric analysis. *Computers and Industrial Engineering*, **113**, pp. 614–629, 2017.

[29] Herrera-Franco, G., Montalván-Burbano, N., Mora-Frank, C. & Moreno-Alcívar, L., Research in petroleum and environment: A bibliometric analysis in South America. *International Journal of Sustainable Development and Planning*, **16**(6), pp. 1109–1116, 2021.

[30] Fahimnia, B., Sarkis, J. & Davarzani, H., Green supply chain management: A review and bibliometric analysis. *International Journal of Production Economics*, **162**, pp. 101–114, 2015.

[31] del Río-Rama, M. de la C., Maldonado-Erazo, C.P., Álvarez-García, J. & Durán-Sánchez, A., Cultural and natural resources in tourism island: Bibliometric mapping. *Sustainability*, **12**(2), p. 724, 2020.

[32] Archambault, É., Campbell, D., Gingras, Y. & Larivière, V., Comparing bibliometric statistics obtained from the Web of Science and Scopus. *Journal of the American Society for Information Science and Technology*, **60**(7), pp. 1320–1326, 2009.

[33] Martín-Martín, A., Thelwall, M., Orduna-Malea, E. & Delgado López-Cózar, E., Google Scholar, Microsoft Academic, Scopus, Dimensions, Web of Science, and OpenCitations' COCI: A multidisciplinary comparison of coverage via citations. *Scientometrics*, **126**(1), pp. 871–906, 2021.

[34] van Eck, N.J. & Waltman, L., Software survey: VOSviewer, a computer program for bibliometric mapping. *Scientometrics*, **84**(2), pp. 523–538, 2010.

[35] Xie, L., Chen, Z., Wang, H., Zheng, C. & Jiang, J., Bibliometric and visualized analysis of scientific publications on Atlantoaxial spine surgery based on Web of Science and VOSviewer. *World Neurosurgery*, **137**, pp. 435–442.e4, 2020.

[36] Herrera-Franco, G., Montalván-Burbano, N., Carrión-Mero, P. & Jaya-Montalvo, M., Worldwide research on geoparks through bibliometric analysis. *Sustainability*, **13**(3), p. 1175, 2021.

[37] Aria, M. & Cuccurullo, C., Bibliometrix : An R-tool for comprehensive science mapping analysis. *Journal of Informetrics*, **11**(4), pp. 959–975, 2017.

[38] Dall'agnol, C.M., de Magalhães, A.M.M., Mano, G.C. de M., Olschowsky, A. & da Silva, F.P., A noção de tarefa nos grupos focais. *Revista Gaúcha de Enfermagem*, **33**(1), pp. 186–190, 2012.

[39] Kinalski, D.D.F., de Paula, C.C., Padoin, S.M. de M., Neves, E.T., Kleinubing, R.E. & Cortes, L.F., Focus group on qualitative research: Experience report. *Revista Brasileira de Enfermagem*, **70**(2), pp. 424–429, 2017.

[40] De Solla Price, D.J. (ed.), *Little Science, Big Science*, Columbia University Press: New York, p. 118, 1963.

[41] Hoffmann, H.K., Sander, K., Brüntrup, M. & Sieber, S., Applying the water–energy–food nexus to the charcoal value chain. *Frontiers in Environmental Science*, **5**, p. 84, 2017.

[42] Yuan, K.-Y., Lin, Y.-C., Chiueh, P.-T. & Lo, S.-L., Spatial optimization of the food, energy, and water nexus: A life cycle assessment-based approach. *Energy Policy*, **119**, pp. 502–514, 2018.

[43] Vanham, D., Does the water footprint concept provide relevant information to address the water–food–energy–ecosystem nexus? *Ecosystem Services*, **17**, pp. 298–307, 2016.

[44] Bellezoni, R.A., Sharma, D., Villela, A.A. & Pereira Junior, A.O., Water–energy–food nexus of sugarcane ethanol production in the state of Goiás, Brazil: An analysis with regional input–output matrix. *Biomass and Bioenergy*, **115**, pp. 108–119, 2018.

[45] Darda, S., Papalas, T. & Zabaniotou, A., Biofuels journey in Europe: Currently the way to low carbon economy sustainability is still a challenge. *Journal of Cleaner Production*, **208**, pp. 575–588, 2019.

[46] Benites Lazaro, L.L., Giatti, L.L. & Puppim de Oliveira, J.A., Water–energy–food nexus approach at the core of business: How businesses in the bioenergy sector in Brazil are responding to integrated challenges? *Journal of Cleaner Production*, **303**, 127102, 2021.

[47] López-Diaz, D.C., Lira-Barragán, L.F., Betzabe González-Campos, J., Serna-González, M., El-Halwagi, M.M. & Ponce-Ortega, J.M., *Optimal Supply Chain for Biofuel Production under the Water–Energy–Food Nexus Framework*, pp. 1903–1908, 2018.

[48] Herrera-Franco, G., Carrión-Mero, P., Montalván-Burbano, N., Mora-Frank, C. & Berrezueta, E., Bibliometric analysis of groundwater's life cycle assessment research. *Water*, **14**(7), p. 1082, 2022.

[49] VanderWilde, C.P. & Newell, J.P., Ecosystem services and life cycle assessment: A bibliometric review. *Resources, Conservation and Recycling*, **169**, 105461, 2021.

[50] Han, X., Zhao, Y., Gao, X., Wang, Y., Jiang, S., Zhu, Y. & An, T., The historical footprint and future challenges of water–energy–food nexus research: a bibliometric review towards sustainable development. *Environmental Reviews*, **29**(2), pp. 260–276, 2021.

[51] Hamidov, A. & Helming, K., Sustainability considerations in water–energy–food nexus research in irrigated agriculture. *Sustainability*, **12**(15), p. 6274, 2020.

WATER–ENERGY–FOOD NEXUS TOOL SELECTION AND APPLICATION OF TOOLS TO A REGIONAL CASE STUDY

FATIMA MANSOUR[1], MAJDI ABOU NAJM[2], ALI YASSINE[3],
ELENA NAJJAR[4] & MAHMOUD AL-HINDI[4]
[1]Department of Civil and Environmental Engineering, American University of Beirut, Lebanon
[2]Department of Land, Air, and Water Resources, University of California, USA
[3]Department of Industrial Engineering and Management, American University of Beirut, Lebanon
[4]Department of Chemical Engineering and Advanced Energy, American University of Beirut, Lebanon

ABSTRACT

Energy, as an integral resource to everyday life, cannot be separated from the other resources involved in its extraction, production, and consumption. The water–energy–food nexus (WEFN) conceptualizes resource sustainability because it frames the interlinkages across sectors, accounting for resource flow across systems. Thus, future sustainable energy resource planning goes hand in hand with overall resource and environmental sustainability. This entails that systems be considered holistically, especially in the planning and production stages, and this can be achieved by applying tools that address the integrated WEFN. While the literature supports a growing number of these tools, the breadth of WEFN applications renders it difficult to determine which tools are suitable for different objectives and users. This work provides a systematic selection approach that applies multi-criteria decision making to determine which tool best fits a set of circumstances. The selection criteria are developed to reflect tool features as comprehensively as possible, accounting for: economic, social, and environmental factors, temporal/spatial scales, stakeholder inclusion, parameter comprehensiveness, and tool complexity, flexibility, and accessibility. Three unique user (scientific user, policymaker, and non-technical stakeholder) scenarios are simulated via distinctive weighting schemes. Of an initial pool of 40 tools, four tools are selected across varied user scenarios. The selected tools are applied to a case study to better demonstrate how the tools differ in performance, as is appropriate for respective user objectives. Case study applications indicate that results are sensitive to assumptions and input data, and that the nature of results across tools differ significantly. The work is versatile in that selection criteria and weighting schemes can be more specifically tailored to cater to specific user needs. Once the WEFN approach is integrated into energy management, the overall perspective on sustainability will improve as advancement in one aspect does not come at the expense of another.

Keywords: water–energy–food nexus, multi-criteria decision making, analytical hierarchy process, resource management, tools, selection, sustainability.

1 INTRODUCTION

With the projected increase of the global population to reach 9.8 billion by 2050, the corresponding increase in resource demand is 55% for water, 60% for food, and 50% for energy [1]–[3]. However, reducing the global demand for a specific resource is not a problem that can be addressed in silo, as this reduction may come at the expense of another resource, possibly exacerbating the problem overall [4]. This is due to the interconnected nature of resources, where water is needed for energy production, energy is needed for water treatment, and both energy and water are needed for food production. The water–energy–food nexus (WEFN) is a concept that accounts for the interlinkages of these resources and thus provides a context for problem solving more sustainably, especially as relates to resource management [5].

The general scientific consensus is that energy consumption has come to be a critical environmental problem that must be addressed. This is because energy is directly related to economic growth, in turn accounting for 87% for global greenhouse gas (GHG) emissions [6]. The situation is especially challenging because, even at this rate of consumption, there

WIT Transactions on Ecology and the Environment, Vol 255, © 2022 WIT Press
www.witpress.com, ISSN 1743-3541 (on-line)
doi:10.2495/EPM220081

remains 1.2 billion people without access to electricity [7], meaning that as living conditions improve, it is expected that GHG emissions will also increase.

To address this problem, it is imperative that energy problems be considered in a WEFN context to prevent the transfer of problems from one resource sector to another and improve overall sustainability [4]. Furthermore, a WEFN framework allows for the incorporation of related challenges (urbanization and economic growth) for more inclusive solutions [8]. An intermediary solution is to leverage existing WEFN tools that cater to sustainable resource allocation and management.

However, while there is a growing number of WEFN tools, little work addresses the selection of a suitable tool for an intended use. While Rosales-Asensio et al. [9] put forth a WEFN decision making framework targeting energy management, they do not apply the suggested criteria to select a tool. Dargin et al. [10] propose a complexity index, developed from eight criteria, to determine the "usability" of eight WEFN tools. Schull et al. [11] apply multi-criteria decision analysis to select a WEFN tool from a total of seven tools using qualitative criteria. In these works, limited tools and criteria are used.

This work puts forth a more involved selection approach, with a versatile nature that allows for its replication with more specific aims in mind. The objectives of this work are: (i) to identify which tools are suitable for differing user needs; (ii) to compare the performance of the selected tools; and (iii) to provide recommendations for further tool improvement. This is accomplished by using multi-criteria decision making (MCDM) to determine which tools best assess the WEFN across different scenarios and applying the selected tools to a specified case study to examine their performance.

2 METHODOLOGY

The applied approach consists of two sections. The first pertains to the WEFN tool selection, and the second applies the selected tools to a case study for further analysis.

2.1 WEFN tool selection

MCDM is applied for the selection approach, because the use of multiple criteria allows for more inclusive assessment of measures that would not be typically included within a single scientific method (due to inability to directly compare these criteria) [12]. In addition, the use of multiple criteria has the capability of addressing diverging goals at the same time [13]. Decision maker preference guides the process to ensure that the most suitable choice is selected [14]. The tool selection approach consists of: (i) identifying the initial set of suitable tools; (ii) specifying the assessment criteria; (iii) rating the tools according to the criteria, iv) generating criteria weights for designated user scenarios; and (iv) applying MCDM to select the tools.

2.1.1 Initial set of WEFN tools

The first step of the selection process is determining the initial set of WEFN tools. These tools are extracted from the literature by applying three filter criteria: (i) a tool must include all three sectors of the WEFN; (ii) the tool must be available; and (iii) tool must be applicable to a case study. Scopus (scientific database), the Nexus Resource Platform website (https://www.water-energy-food.org/), and literature reviews addressing nexus tools [15]–[18] were studied to identify potential tools. Based on these resources and in accordance with the filter criteria, a total of 40 tools are initially selected (as of December 2020).

2.1.2 Criteria development

A set of ten criteria is developed to cover WEFN tool characterization, based on a study of the WEFN and MCDM literature [7], [13], [16], [18]–[22]. The criteria span different aspects of tool use and endeavour to account for generic considerations that are made when selecting tool, going beyond the scientific aspects. The criteria catering to the scientific aspects of tool use are: *factors* (referring to the number of economic, environmental, and social factors included in the tool), *parameter comprehensiveness* (i.e. the number of parameters capturing nexus interlinkages), *scenario building* (reflecting tool's capacity for additional sensitivity analyses and scenario changes), *spatial scales* (how applicable the tool is across areas: local/ national/regional/global), *temporal scales* (whether the tool includes time-related considerations in its analyses), and *uncertainty* (if the tool accounts for risk and associated uncertainty). The non-scientific criteria are: *availability* (how easily a user can access the tool), *model complexity* (as a proxy for tool's ease of use), *flexibility* (how easy it is to adapt the tool to other case studies), and *stakeholder inclusion* (to assess how well the tool caters to other related parties).

The tools are assessed based on how well they rate with respect to these criteria. The rating is based on a rubric that defines scores numerically and qualitatively for maximal transparency. The rubric is defined such that each criterion has up to five levels corresponding to a specific numerical score (ranging from 1 to 9, the latter being the most preferred). For example, if a tool can only be applied to one specific city, it is assigned a score of 1; whereas, if it can be applied across multiple (two or more) spatial scales, it is assigned a score of 9. The same concept is applied to all criteria, noting that other criteria might have more than two levels. The factors criterion is a special case, such that the score is the average of scores for each of economic, environmental, and social factors separately, such that the scores are: 1 if no factors are included, 3 if one factor is included, 5 if two factors are included, 7 if three factors are included, and 9 if more than three factors are included. The scores for each factor type (economic, environmental, and social) are averaged for a final score for the factors criterion.

2.1.3 Criteria weighting

Three alternative weighting schemes, in addition to an equal weighting scenario, are applied to model tool selection for three different users: (i) scientific user; (ii) policymaker; and (iii) non-technical stakeholder. Weights are generated to reflect diverse user preferences, which may render a different tool more suitable for each scenario. The SMARTER (Simple Multi-Attribute Rating Technique Exploiting Ranks) method is used to determine criteria weights. SMARTER operates based on order of importance ranking, where the decision maker ranks the criteria set in order of consecutive importance, and this order is converted into a set of weights using the rank order centroid [23], using Logical Decisions v7.2 software.

2.1.4 Decision matrix

Logical Decisions v7.2 software is also used for the selection analysis, requiring the input of the decision matrix. The decision matrix is compiled from: (i) WEFN tools as alternatives; (ii) developed criteria as assessment measures; and (iii) the tool scores; and (iv) the criteria weights for each user scenario. The overall performance of an alternative is calculated with a linear additive model that sums the product of each criterion weight and corresponding rating.

2.2 Case study on selected tools

The selected tools comprise those ranking first in each of the scenarios described in the previous section. A case study is applied to each of these tools to analyse how and why results may differ. The objective of the case study is to assess the Lebanese WEFN. As such, a point of reference, resource consumption (for the year 2018), is specified to enable numerical comparison across different tools.

Lebanon is located in the Mediterranean region, with a population of 6.85 million [24]. Lebanon imports over 97% of its energy mostly in the form of petroleum [2]. Approximately 48% of it is used for electricity production (generating approximately 21.24 GWh [2]), the remaining is used for non-energy functions. In terms of water resources, Lebanon primarily relies on surface water and groundwater, consuming approximately 1500 million cubic meters of water [25]. Lebanon's food sector consists of livestock and agricultural crops. For the comparison, tool performance is gauged by how accurate simulated resource (crops, water, and energy) consumption values are compared to actual resource consumption (derived from global and national databases). It is noted that certain assumptions are made with respect to consumption values, as relates to domestic supply, and based on available data.

Fig. 1 summarizes the proposed methodology.

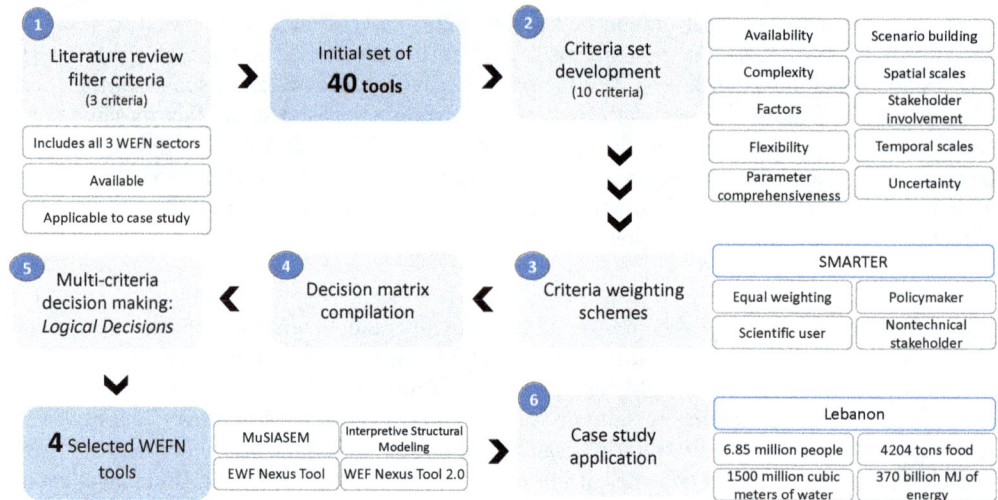

Figure 1: Summary of selection approach methodology.

3 RESULTS

The first set of results pertains to the WEFN tool selection, while the second presents the results of tool application to the described case study.

3.1 WEFN tool selection

The weighting schemes generated for each user scenario are summarized in Table 1. Parameter comprehensiveness and scenario building are the most important criteria for the scientific user scenario. On the other hand, stakeholder inclusion and the factors criteria are

Table 1: Criteria weights for three user scenarios.

Criteria	User weighting schemes		
	Scientific user	Policymaker	Non-technical stakeholder
Availability	0.034	0.048	0.193
Complexity	0.021	0.143	0.293
Factors	0.048	0.193	0.085
Flexibility	0.085	0.034	0.065
Parameter comprehensiveness	0.293	0.085	0.110
Scenario building	0.193	0.110	0.143
Spatial scales	0.065	0.001	0.048
Stakeholder inclusion	0.010	0.293	0.010
Temporal scales	0.110	0.065	0.034
Uncertainty	0.143	0.021	0.021

the more important ones for the policymaker user. Tool complexity and availability are the driving criteria for the non-technical stakeholder.

The tools selected for the designated scenarios are presented in Table 2.

Table 2: Selected tools for baseline and user scenarios.

User scenario	Selected tool	Description of tool
Baseline	MuSIASEM [26]	This tool, Multi-Scale Integrated Analysis of Societal and Ecosystem Metabolism (MuSIASEM), applies a resource accounting method that studies metabolic patterns in society. It analyses data at different hierarchal scales allowing for better analysis of nexus synergies (energy, food, and water and their interrelations) and consequent societal impacts.
Scientific user	EWF Nexus Tool [27]	This tool is based on life cycle analysis (LCA) and assesses the performance of WEF systems from the cradle to grave via the quantification of environmental impacts. Each nexus resource has a subsystem, such that transfers between these subsystems represent the nexus interactions
Policy maker	ISM [28]	This tool, Interpretive Structural Modelling (ISM), converts nexus interlinkages and interdependencies into a well-defined model using the concepts of reachability and transitive inference. The tool results in a hierarchy of WEFN factors using feedback and input from a panel of experts/ stakeholders.
Non-technical stakeholder	WEF Nexus Tool 2.0 [29]	This tool uses a scenario-based framework that quantifies flows of matter and energy between nexus dimensions. Input and output vales are then used to compute an overall sustainability index, which is a decision-making indicator for comparison between different scenarios.

The four tools selected across scenarios score highly with respect to the key criteria characterizing each scenario. The MuSIASEM tool is the one chosen for the baseline. Its scoring shows it to be a good tool in terms of covering most criteria bases. The EWF Nexus Tool was selected for the scientific user scenario because it caters to extensive technical modelling with very detailed data requirements and comprehensive accounting for embedded processes. As for the policymaker scenario, the Interpretive Structural Modelling (ISM) tool directly integrates expert/stakeholder dialogues (in the form of surveys or panels), rendering it a suitable choice for this scenario. The WEF Nexus Tool 2.0 is suitable for the nontechnical stakeholder, as it is user friendly and can be used without any prior training.

3.2 Case study on selected tools

To apply to the case study of Lebanon, the required input data is adjusted accordingly for the different tools. Table 3 presents a summary of the resulting WEFN resource consumption across the selected tools, as applicable.

Table 3: Summary of total resource consumption across tools.

Tools	Resources consumed		
	Water (m^3)	Energy (MJ)	Food (10^6 tons)
EWF Nexus Tool	1.45E+08	3.68E+11	2.52
ISM	5.96 E+08	3.70E+11	4.20
MuSIASEM	5.96 E+08	3.70E+11	4.20
WEF Nexus Tool 2.0	0.08 E+03	5.86E+08	–

The application of MuSIASEM entails obtaining the overall resource consumption values from the appropriate databases and breaking down across a specified societal hierarchy (of sectors). On the other hand, ISM does not include direct quantitative values. The tool generates a hierarchy of factors indicating which are most important for the assessed WEFN context. The tabulated values are simply attributed to the consumption factors post-analysis. In other words, water, energy, and food consumption are the factors included in the ISM analysis (and within the generated factor hierarchy), but the values in the table are those obtained from the literature for Lebanon and simply assigned to the considered factors as appropriate for the purpose of numerical comparison. As such, given the more involved nature of the tool, for the objective of this comparison and using existing databases as a reference, MuSIASEM is deemed most accurate in capturing WEFN interactions.

The EWF Nexus Tool results are within the same range of magnitude as those of MuSIASEM. This is also because the EWF Nexus Tool employs researched data values specific to the Lebanese WEFN. The differences between the values generated by these two tools are due to: (i) differing assumptions made based on available data; and (ii) embedded processes within life cycle analysis (LCA) that are not accounted for in MuSIASEM. For example, in the application of the EWF Nexus Tool, only a limited number of crops were included in the simulation and water consumption was limited to irrigation and electricity production. In addition, the LCA simulation of a crop includes fertilizer and pesticide use, which are not typically included in generic consumption values. The ability of these two tools to take on specific context-based values renders them more accurate than the remaining tools.

With respect to the WEF Nexus Tool 2.0, although it does incorporate WEFN resources, results are specific to the food production sector. Accordingly, the water consumed is less than that in MuSIASEM because the water considered in the tool is only that used for specific crop production. On the same note, the energy for food production in the WEF Nexus Tool is lower than the total amount of energy consumed in Lebanon as a whole (before energy sector consumption). The consumed amount is still high considering that only the food production sector is assessed. This can be explained by the tool's inclusion of energy consumption in irrigation, other agriculture-related process, and transport, which are not explicitly included in the other tools and are designed to cater to Qatar's specifications. It is noted that the crops available in the WEF Nexus Tool 2.0 do not capture those that are important to the Lebanese food sector, rendering the results incomplete.

4 DISCUSSION

The first section of discussion addresses the selected tools for the simulated users, the results of the case study, and the recommendations for a "universal" tool. The second section of the discussion examines the versatility of the applied selection approach, focusing on how the approach can be adapted to serve other users and more specific objectives.

Before delving into these discussions, the suggested selection approach is compared to the limited work done around this topic. Rosales-Asensio et al. [9] uses three criteria (two out of three nexus sectors, capability of evaluation at high level, and open accessibility and availability) to determine the initial set of tools to which three examination principles are to be applied: input entry requirements (in terms of data), output expectations (i.e., what inquiries can be answered with the tool), and analytical aspects of the tool. However, tool selection is not effectively applied to showcase the framework's capabilities. In the determination of their complexity index, Dargin et al. [10] use eight criteria (data accessibility, user friendliness, accessibility, data granularity, data requirements, training intensity, subject matter expertise requirement, and user defined scenarios). The computed index is meant to be used as a decision aid for tool comparison and selection with respect to nexus scope and externalities, tool methodologies, and tool purposes. Schull et al. [11] focus on the selection of water management-oriented tools within a nexus context. The initial set of tools was obtained from the literature using an exploratory approach. The criteria applied to the set of tools are: availability and accessibility, user friendliness and simplicity, flexibility, comprehensiveness, and predictive component. An unweighted qualitative decision matrix approach is used to rate the tools for selection.

In this work, the selection approach is meant to identify the suitable WEFN tool for a specific user. Compared to the other selection approaches, this work considers a larger initial pool of tools identified using three distinct filter criteria, permitting a wider range of choice. In addition, a larger number (ten) of evaluation criteria are applied, with detailed descriptions of the criteria and their levels for more specific rating and overall utility calculation. The criteria also span a broader range of tool aspects for more comprehensive tool characterization, both technically (ex. spatial and temporal scales) and beyond the technical (ex. stakeholder inclusion). The approach also ties in user preference with respect to the expected objective behind tool use to guide the selection process by simulating two potential users and correspondingly deriving the appropriate weighting schemes. Furthermore, a comparative analysis is conducted by applying the same case study to the four tools selected across the varying user scenarios, which to the authors' knowledge has not been done before. The application of the same case study to different tools allows for deeper insight to the analytical differences of tools and provides a basis for fit analysis with respect to user needs.

4.1 Selected tools, case study applications, and "universal" tool specifications

More accurate results tend to come at the cost of data extensiveness and time-consuming applications. As such, MuSIASEM tool is relatively effective at characterizing the WEFN due to its detailed (level of detail is user specified) flow of elements within the system, and its relatively straightforward application. It enables the user to identify sources and sinks in the system, thus highlighting sectors of high consumption and determining the level of dependency among resources. The tool is more of an open-ended methodology that can be customized to context specifications. As for the EWF Nexus Tool, aside from the ability to specify production amounts and detailed input factors, the software simulates resource flows across specified processes and includes all embedded processes in the simulation. Additionally, this tool calculates the environmental impacts associated with the simulated processes. However, the EWF Nexus Tool is difficult to use and requires a significant amount of time to learn software application and acquire data.

While the ISM tool does provide a qualitative analysis in terms of factor importance, it does not provide any information about the magnitude of flows between sectors, and consequently cannot effectively model the WEFN. It is useful in that the hierarchical structure can provide direction in terms of what factors and/or sectors should be focused on. However, it is highly subjective to expert judgment in the formation of connections. As for the WEF Nexus Tool 2.0, it has a limited set of options for inputs and outputs, with major assumptions that significantly impact results. As such, it cannot be used for accurate modelling of the WEFN in countries other than Qatar. And even then, the analysis is of a very specific nature, catering to food security with a limited selection of crops included.

It is noted that these results are not intended to fully capture the breadth of utilization for these tools, but to provide insight into the performance of the tool in assessing the WEFN. Each tool has its advantages and disadvantages, and further reinforces the difficulty in capturing the WEFN in its entirety and potentially explains why no there is no "universal" WEFN tool.

Perhaps, an alternative to achieving a "universal" WEFN tool is combining existing tools so that they compensate for what the other lacks. Based on the selected tools in this work, an all-around tool would combine the key contributions from each tool and allow the user to select the level of modelling complexity and system specificity needed. It would: (i) integrate stakeholder involvement (ISM and MuSIASEM); (ii) account for environmental impacts (EWF Nexus Tool); (iii) characterize WEFN resource flows across multiple levels, including climatic zones (MuSIASEM and EWF Nexus Tool); (iv) identify factors of strength and liability (ISM); (v) account for trade flows (WEF Nexus Tool 2.0, MuSIASEM, and EWF Nexus Tool); and (vi) have a user-friendly interface (WEF Nexus Tool 2.0).

4.2 Selection approach versatility

The vastness of the nexus, its many interconnections, and potentially diverging stakeholder interests render it nearly impossible to satisfy all user interests simultaneously, particularly with a single tool. This is beside the conflicting nature of some criteria. For example, it is the general rule that more accurate and comprehensive tools (characterized by high ratings on: *parameter comprehensiveness, spatial scales, temporal scales,* and *uncertainty*) tend to require more data for application and are harder to use (*model complexity* and *availability*). As such, it is worth investigating approaches that cater to the selection of a WEFN tool as suits a particular user.

In the presented approach, the developed selection criteria set spans multiples bases of interest, and then proposes weighting in accordance with user interest. In cases where the user has a very specific objective in mind, more criteria can be added, or existing ones parsed out to better reflect the characteristics the user is looking for in a tool. For example, if a user wants to focus on energy production, a filter criterion can be added where the initial pool of tools consists of those that focus on energy in a WEFN context. With that in hand, more specific criteria can be applied as measures to select the most suitable tool. Building on the previous example, tailoring this approach to energy production, the criteria set can be customized include more technical factors such as: specific energy costs, land requirements for energy technologies, and water requirements for energy production. A rubric can be constructed with levels for each criterion based on number of energy technologies or cost factors included.

As for the weight derivation process, it functions to accommodate user preference so that the selected tool best reflects user interest. With respect to the simulated user scenarios, generic user roles are assumed for demonstrative purposes. They are not meant to capture the depth of these roles. However, they provide an example of how criteria importance differs from one user to another. Adapting criteria weights, whether it be via the same weight generation method or another, allows the user to determine the importance of each criterion as best fits a particular scenario or reflects a user's background.

This is not to say that a tool will be found to cater to every need. In such a case, an apparent literature gap is highlighted, and further work can be done that basis.

5 CONCLUSION

The WEFN, in its conceptual nature, deals with resource production, management, and allocation under the umbrella of environmental sustainability. Naturally, tools developed in line with this premise will also cater to these issues. However, the breadth of the WEFN in its spanning all three resource sectors, accounting for interlinkages between these sectors, and expanding beyond the scientific domain for effective implementation gives rise to a myriad of tools with dissimilar, albeit WEFN related, objectives. The abundance of nexus tools, with no systematic selection strategy, complicates an intrinsically complicated problem. As no single nexus model can address all WEFN aspects, as of yet [30], moving forward, it may be worth to focus on developing more advanced selection processes so as to provide a systematic means for determining which WEFN tool best applies to a given situation or specified set of user preference.

This work proposes an MCDM selection strategy designed to cater to different user needs (scientific user, policymaker, non-technical stakeholder). The selected tools are MuSIASEM (baseline scenario with equally weighted criteria), EWF Nexus Tool (scientific user scenario), ISM (policymaker user scenario), and the WEF Nexus Tool 2.0 (non-technical stakeholder scenario). The performance of the selected tools varies in accordance with varying user needs. Case study applications of these tools to the Lebanese WEFN show how each tool differs in terms of input, methodology, and output. Beyond the demonstrative scope of this work, it is noted that all tools can be used, with different objectives, for more (or less) detailed analysis, as per the user's needs.

The presented selection analysis provides a rudimentary rundown of existing WEFN tools for different use scenarios. With the growing number of tools in the literature, the initial set of tools can be further expanded. The tool selection approach caters to environmental sustainability by promoting WEFN tool utilization and ensuring that improvement in one environmental compartment does not come at the expense of another. However, a key feature of this work is its versatility in fitting to more specific environmental objectives such that the specified objective is addressed with a suitable tool in a WEFN context. Tailoring the selection process can come in the form of more specific criteria, a modified scoring rubric, and more particular weighting schemes, all in accordance with the user's objectives. Future improvements to the approach include: (i) incorporation of the growing number of WEFN tools; (ii) more specific tool use objectives so that outputs can be aligned with specific tool uses/expectations/outputs; and (iii) the incorporation of expert opinion to the weight derivation process to better cater to appropriate tool use.

ACKNOWLEDGEMENT
The authors would like to thank the Maroun Semaan Faculty of Engineering and Architecture at the American University of Beirut for making this research possible.

REFERENCES
[1]   Boretti, A. & Rosa, L., Reassessing the projections of the World Water Development Report. *npj Clean Water*, **2**(1), 2019. DOI: 10.1038/s41545-019-0039-9.
[2]   IEA, Data and statistics. https://www.iea.org/data-and-statistics/data-tables?country= LEBANON.
[3]   EIA, EIA projects nearly 50% increase in world energy usage by 2050, led by growth in Asia, Energy Information Administration. https://www.eia.gov/todayinenergy/detail.php?id=41433.
[4]   Liu, J. et al., Nexus approaches to global sustainable development. *Nature Sustainability*, **1**(9), pp. 466–476, 2018. DOI: 10.1038/s41893-018-0135-8.
[5]   Ibrahim, M.D., Ferreira, D.C., Daneshvar, S. & Marques, R.C., Transnational resource generativity: Efficiency analysis and target setting of water, energy, land, and food nexus for OECD countries. *Science of the Total Environment*, **697**, 134017, 2019. DOI: 10.1016/j.scitotenv.2019.134017.
[6]   Roser, M., The world's energy problem. Our World in Data. https://ourworldindata.org/worlds-energy-problem.
[7]   Keairns, D.L., Darton, R.C. & Irabien, A., The energy-water-food nexus. *Annual Review of Chemical and Biomolecular Engineering*, **7**, pp. 239–262, 2016. DOI: 10.1146/annurev-chembioeng-080615-033539.
[8]   FAO, Water-energy-food nexus for the review of SDG 7, 2018.
[9]   Rosales-Asensio, E., de la Puente-Gil, Á., García-Moya, F.-J., Blanes-Peiró, J. & de Simón-Martín, M., Decision-making tools for sustainable planning and conceptual framework for the energy–water–food nexus. *Energy Reports*, **6**, pp. 4–15, 2020. DOI: 10.1016/j.egyr.2020.08.020.
[10] Dargin, J., Daher, B. & Mohtar, R.H., Complexity versus simplicity in water energy food nexus (WEF) assessment tools. *Science of the Total Environment*, **650**, pp. 1566–1575, 2019. DOI: 10.1016/j.scitotenv.2018.09.080.
[11] Schull, V.Z., Daher, B., Gitau, M.W., Mehan, S. & Flanagan, D.C., Analyzing FEW nexus modeling tools for water resources decision-making and management applications. *Food and Bioproducts Processing*, **119**, pp. 108–124, 2020. DOI: 10.1016/j.fbp.2019.10.011.
[12] Diaz-Balteiro, L., González-Pachón, J. & Romero, C., Measuring systems sustainability with multi-criteria methods: A critical review. *European Journal of Operational Research*, **258**(2), pp. 607–616, 2017. DOI: 10.1016/j.ejor.2016.08.075.
[13] Wątróbski, J., Jankowski, J., Ziemba, P., Karczmarczyk, A. & Zioło, M., Generalised framework for multi-criteria method selection. *Omega*, **86**, pp. 107–124, 2019. DOI: 10.1016/j.omega.2018.07.004.

[14] Cinelli, M., Coles, S.R. & Kirwan, K., Analysis of the potentials of multi criteria decision analysis methods to conduct sustainability assessment. *Ecological Indicators*, **46**, pp. 138–148, 2014. DOI: 10.1016/j.ecolind.2014.06.011.

[15] Dai, J. et al., Water-energy nexus: A review of methods and tools for macro-assessment. *Applied Energy,* **210**, pp. 393–408, 2018. DOI: 10.1016/j.apenergy.2017.08.243.

[16] Kaddoura, S. & El Khatib, S., Review of water-energy-food Nexus tools to improve the Nexus modelling approach for integrated policy making. *Environmental Science and Policy*, **77**, pp. 114–121, 2017. DOI: 10.1016/j.envsci.2017.07.007.

[17] Mannan, M., Al-Ansari, T., Mackey, H.R. & Al-Ghamdi, S.G., Quantifying the energy, water and food nexus: A review of the latest developments based on life-cycle assessment. *Journal of Cleaner Production*, **193**, pp. 300–314, 2018. DOI: 10.1016/j.jclepro.2018.05.050.

[18] Al-Saidi, M. & Elagib, N.A., Towards understanding the integrative approach of the water, energy and food nexus. *Science of the Total Environment*, **574**, pp. 1131–1139, 2017. DOI: 10.1016/j.scitotenv.2016.09.046.

[19] Fayiah, M., Dong, S., Singh, S. & Kwaku, E.A., A review of water–energy nexus trend, methods, challenges and future prospects. *International Journal of Energy and Water Resources*, **4**(1), pp. 91–107, 2020. DOI: 10.1007/s42108-020-00057-6.

[20] Shannak, S.d., Mabrey, D. & Vittorio, M., Moving from theory to practice in the water–energy–food nexus: An evaluation of existing models and frameworks. *Water-Energy Nexus*, **1**(1), pp. 17–25, 2018. DOI: 10.1016/j.wen.2018.04.001.

[21] Zhang, P. et al., Food-energy-water (FEW) nexus for urban sustainability: A comprehensive review. *Resources, Conservation and Recycling*, **142**, pp. 215–224, 2019. DOI: 10.1016/j.resconrec.2018.11.018.

[22] Purwanto, A., Sušnik, J., Suryadi, F.X. & de Fraiture, C., Water-energy-food nexus: Critical review, practical applications, and prospects for future research. *Sustainability*, **13**(4), 2021. DOI: 10.3390/su13041919.

[23] Riabacke, M., Danielson, M. & Ekenberg, L., State-of-the-art prescriptive criteria weight elicitation. *Advances in Decision Sciences*, **2012**, pp. 1–24, 2012. DOI: 10.1155/2012/276584.

[24] World Bank, Lebanon. https://data.worldbank.org/country/LB.

[25] Fanack, W., Water resources in Lebanon. https://water.fanack.com/lebanon/water-resources/.

[26] Giampietro, M. et al., An innovative accounting framework for the food-energy-water nexus: Application of the MuSIASEM approach to three case studies. Food and Agriculture Organization, 2013.

[27] Al-Ansari, T., Korre, A., Nie, Z. & Shah, N., Development of a life cycle assessment model for the analysis of the energy, water and food nexus. *24th European Symposium on Computer Aided Process Engineering*, pp. 1039–1044, 2014.

[28] Li, G., Huang, D., Sun, C. & Li, Y., Developing interpretive structural modeling based on factor analysis for the water-energy-food nexus conundrum. *Science of the Total Environment*, **651**, pp. 309–322, 2019. DOI: 10.1016/j.scitotenv.2018.09.188.

[29] Daher, B.T. & Mohtar, R.H., Water–energy–food (WEF) Nexus Tool 2.0: Guiding integrative resource planning and decision-making. *Water International*, **40**(5–6), pp. 748–771, 2015. DOI: 10.1080/02508060.2015.1074148.

[30] McCallum, I. et al., Developing food, water and energy nexus workflows. *International Journal of Digital Earth*, **13**(2), pp. 299–308, 2019. DOI: 10.1080/17538947.2019.1626921.

ENERGY EFFICIENCY THROUGH WATER-USE EFFICIENCY IN LEISURE CENTRES

AISHA BELLO-DAMBATTA & PRYSOR WILLIAMS
School of Natural Science, Bangor University, UK

ABSTRACT

Climate change poses significant challenges, and the global community is not on track to meet sustainable development goals or the Paris Agreement to mitigate climate change. The COVID-19 pandemic and necessary government measures to curb the spread of the virus has put climate action on hold and shut down economies. The need for improved ventilation as an important mitigating factor against the risk of COVID-19 transmission has additional implications for costs and emissions for businesses. Leisure centres, as large users of water and energy, account for significant emissions and operational costs. However, there is scope for significant reductions in water and water-related energy demands and associated emissions and costs without impacting service quality and delivery. These reductions can be a promising response to the current challenges of climate change and post-COVID-19 economic recovery, particularly given current UK energy crises and inflation trends. We have been working with leisure centres to support them in improving energy efficiency through water-use efficiency as part of the cross-border, interdisciplinary Interreg Dŵr Uisce research project on improving the energy performance and long-term sustainability of the water sectors in Ireland and Wales. In this paper, we discuss the potential of energy efficiency gains based on the framework on water management hierarchy which prioritises management actions in order of preference of implementation, where the next hierarchy should only be considered once all potential savings from the hierarchy above have been exhausted. We also discuss how these interventions are not one-size-fits-all – although leisure centres typically have the same water-use types, they differ significantly in age, size, location, building types and materials, functionality, and efficiency; and why therefore, interventions must be considered on a site-specific and case-by-case basis.
Keywords: climate action, energy efficiency, water efficiency, water–energy nexus, heat recovery, sustainability.

1 INTRODUCTION

The water industry is highly energy intensive and on average between 2% and 3% of the world's energy use is used to treat water to potable quality, deliver it to consumers, and to process and dispose of wastewater. In the UK, for example, up to 3% of total energy consumption is by water companies [1]. However, this represents only around 11% of actual water-related energy consumption, with most of the water-related energy use attributed to water demand [2]. Given this obvious link between water and energy use, reducing the water for demand, especially in hot water use, can significantly reduce water-related energy demand and associated GHG emissions and costs.

Better management of water demand is promising as both a climate change mitigation and adaptation strategy, and can reduce water consumption to conserve the resource, thus reducing the energy need and associated GHG emissions, operational costs, and to reduce negative environmental impacts without impacting service quality and delivery. This also makes business sense as even the simplest interventions can provide significant savings in operational costs and environmental taxes. In the UK for example, some local authorities or business may even be able to claim capital allowances for investing in low-carbon technology through schemes like Non-Domestic Renewable Heat Incentive and water efficient enhanced capital allowances. It is good practice to also consider water and energy management as a single, integrated management task rather than two separate tasks as is currently typically the

WIT Transactions on Ecology and the Environment, Vol 255, © 2022 WIT Press
www.witpress.com, ISSN 1743-3541 (on-line)
doi:10.2495/EPM220091

case because one typically affects the other. This is in line with both the UK and Welsh governments' ambitious targets to reduce GHG emissions to net zero by 2050 and other initiatives such as the Welsh governments' Well-being of Future Generations (Wales) Act 2015, the UN Sustainable Development Goals, and the European Green Deal, therefore water management should be considered as a core component of energy management policy and carbon reduction ambitions and targets.

Leisure centres are large users of water and energy and account for significant emissions and operational costs. Progress is being made in reducing energy use in leisure centres, particularly in Local Authority (LA) owned centres. All the LA-owned leisure centres we worked with have made considerable investments and improvements in energy management, however, not much has been done in terms of water management except from ongoing and phased refurbishment of water using fittings and appliances even with the considerable scope for very significant reductions in water and water-related energy demands and associated emissions and costs without impacting service quality and delivery. These reductions can be a promising response to the current challenges of climate change, post-COVID-19 economic recovery, and the current UK energy prices.

It is important to also consider the whole building in leisure centres and not just their Mechanical, Electrical, and Plumbing (MEP) systems as a more cost-effective way to improve energy efficiency, and adoption of new standards like Passive House [3] and Part L of the Building Regulations [4] and technical guidance from industry groups such as Pool Water Treatment Advisory Group (PWTAG) [5] will also lead to much better energy efficiencies. Policy is a very important starting point to help drive change and reasonably moving in the right direction in terms of new-builds as it is much easier to do this with new-builds that must go through the regulatory and compliance process than it is with retrofitting existing leisure centres that form most of the leisure centres in the UK, even with the quiet reasonable impact it can have on cost and carbon savings. One way leisure centres can take initiative and positive climate action and have a critical mass emissions reduction as a sector is to work very closely with industry (e.g., through PWTAG) to embed the idea of reducing carbon footprint into operation guidelines.

2 WATER AND WATER-RELATED ENERGY USE IN LEISURE CENTRES

To identify areas with potential water–energy savings in leisure centres, it is important to understand how water is used in the centres and how this relates to energy use. It is estimated that water, energy, and waste costs can account for over 30% of the running costs in leisure centres with swimming pools [6], [7].

Research indicates that in general swimming pool and pool hall; Heating, Ventilation, and A/C (HVAC) systems; and domestic hot water heating represent the highest consumption of water and energy in leisure centres [7]. The Carbon Trust also identifies pool hall, ventilation and A/C systems, and space heating to be the common areas where energy is typically wasted [6]. This provided a baseline indication of the areas of focus at the leisure centres we audited in Wales. In general, the audit process was used to assess:

- the types of wet leisure centres;
- how water is used in the different leisure centres;
- the areas where water and water-related energy savings and emissions can be made;
- how this can be achieved;
- what this can mean in terms of emissions and costs savings; and
- how this can be used to improve the carbon footprint and wider environmental performance of the leisure centres.

The audits allowed for the benchmarking of water and water-related energy use the centres and a better understanding of where water-related energy and emissions savings can be made. The managers and operators of the centres interviewed during the audit process recognised the need for adopting a water management policy and strategy and had some knowledge and understanding of how this can be achieved in practice. The audits found the leisure centres to be generally water efficient – in all but a few centres which had issues with leaking WCs, dripping taps, and push taps no longer performing effectively, particularly, but not exclusively with the oldest fittings.

Some of our recommendations on the ways leisure centres can reduce their energy use through water-use efficiency, include metering and regular monitoring of water consumption, regular site walk-arounds and continuous record keeping of water use and water fittings and appliances, implementing a programme of interventions focussed on behavioural change in water use, and if possible increasing the stock of low-carbon and/or renewable technology installations at the leisure centres like Wastewater Heat Recovery (WWHR), solar PV, biomass boilers, and heat pumps to further reduce carbon footprint and improve environmental performance.

The recommendations are based on the framework of water management hierarchy (Fig. 1) which prioritises water management actions in order of preference of implementation, where the next hierarchy should only be considered once all potential savings from the hierarchy above have been exhausted. This is based on the principles of waste management hierarchy of the EU Waste Framework Directive [8] which ranks interventions according to their environmental performance with respect to climate change, air and water quality, and resource depletion.

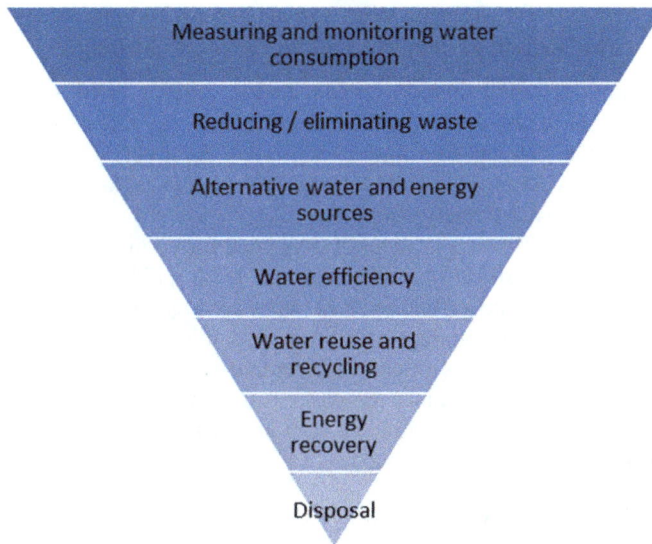

Figure 1: Water–energy management hierarchy (after water management hierarchy).

However, it is important to note that not all interventions will result in reduction in both water and energy use, and some interventions can increase water or energy use. In general, if water or energy demand is to be reduced, then some other factor must change to accommodate this reduction, and where a reduction of water or energy results in an increase of the other, it may not necessarily be clear which is the more sustainable outcome, and this should be decided on project- or site-specific basis [2]. There are also some health and safety implications in water and energy efficiency that must be considered and addressed. Another important note is that these interventions are not one-size-fits-all because although leisure centres typically have the same water-use types, they differ significantly in age, size, location, building types and materials, functionality, and efficiency, and therefore interventions must be considered on a site-specific and case-by-case basis.

It is estimated that leisure centres can save up to a third of heating costs through simple energy efficiency interventions, such as using energy efficient lighting; maintaining appropriate temperatures; ensuring HVAC systems match building occupancy; and maintaining boilers and pipe work [9]. The amount of water and water-related energy demands that can be saved depends on several factors, including building type and age, building materials, specific centre functions, footfall, etc., and these savings are also influenced by the installed water using fittings and appliances, number of people using water on site, the efficiency of the HVAC system, energy source, and system loses [10].

Leisure centres must also meet certain minimum water and energy performance standards for customer comfort, hygiene, and health and safety. Even so, many centres use more water than they need to, leading to higher than necessary water use, sewerage discharge, energy use, emissions, and costs. The leisure centres we audited differed significantly, but they all had typical water-related energy use in these areas:

- plant room (boilers and hot water heaters);
- swimming pool and pool hall (pool heating and HVAC systems);
- domestic hot water use for showers, WCs, urinals, taps in both dry and wet areas, kitchen and cafeteria areas, staff areas, laundry, and for facilities cleaning.

Some of the centres also had shared premises and/or facilities with schools and/or other tenants and some had some low-carbon and/or renewable energy technology such as biomass boilers, heat pumps, Mechanical Ventilation with Heat Recovery (MVHR), and Building Management System (BMS). Some of the leisure centres also replaced all lighting to LEDs and have switched to renewable energy supply.

Of course, with COVID-19 there is an additional need for ventilation and therefore energy use and costs. There is legislation and guidance [11], [12] in relation to minimum requirements for ventilation and efficiencies for heat recovery systems to allow centres to reopen and keep the pool environment as safe as possible. However, the requirement to increase the fresh air and flush out as much of the contaminated air as possible without putting it through the recycling system back into the space is not energy efficient. There is therefore a trade-off between the efficiency gains in carbon savings and the need for additional ventilation to ensure safety.

3 WATER–ENERGY SAVINGS OPPORTUNITIES IN LEISURE CENTRES

By far, swimming pools and pool halls were the largest users of both water and water-related energy use in all the centres, which is in line with estimates that pool heating and ventilation alone can account for up to 65% of energy use – mainly from pumping, filter backwashing, and heating pools and pool halls [7]. Pool temperatures vary depending on activity or use but are typically heated from average mains temperatures which changes seasonally to about 30°C on average. The pool halls were also kept at 1°C above pool temperature to limit evaporation from pool surface.

Pool halls also need to be ventilated to maintain pool hall temperatures both for comfort and to prevent condensation from humidity and damage to building and equipment [9], and HVAC and MVHR systems are highly energy intensive. Energy and money are wasted when air is mechanically removed and replaced by cold air from the outside which needs to be heated or cooled to match inside temperatures [7].

After accounting for swimming pools and pool halls, the largest single factor determining water use in leisure centres is occupancy, which includes both staff numbers and customer footfall. This is because water use is a function of both installed water using fittings and appliances and of water-use behaviours. In general, three variables determine the amount of domestic water use [13]:

- flow rate;
- water-use event duration; and
- frequency of use of water fitting or appliance.

Of these variables, event duration and frequency of use are dependent on user behaviour. Flow rates vary widely and are largely governed by several physical factors, including water pressure and the design of installed fittings or appliances. The frequency of use of use of fittings and appliances is related to occupancy, and in general, the more footfall a centre has, the more water it will use.

3.1 Water-related energy efficiency

One way of reducing water and water-related energy demands in leisure centres is through water-use efficiency, which is often a simple, zero or low-cost intervention with immediate results and short payback periods that can have an important role to play in reducing energy use as well as help reduce operational and maintenance costs whilst still maintaining minimum standards of hygiene and comfort.

Water efficiency can also be used to improve energy performance of leisure centres and their carbon footprint, and improve environmental ratings (e.g., LEED and BREEAM) and therefore overall sustainability and wider environmental performance. It can also be used to reduce the amount and flow pattern of wastewater to drainage systems, resulting in reduction in total volume of wastewater discharge and sewerage costs [13].

Water efficiency can therefore have an important role to play in helping leisure centres meet their carbon targets and responsibilities, as well as help reduce operational and maintenance costs – reducing consumption helps reduce supply costs, increase useful life of water using fittings and devices, which at the very least can result in deferment of investment in upgrades and replacement by reducing the risk of damage to buildings from condensation due to overheating and evaporation. This can be accompanied by fixing, retrofitting, and/or replacing older inefficient or broken water using fittings and appliances, implementing process and operational improvements, and/or social interventions such as behavioural change initiatives or education and awareness raising campaigns for both staff and customers.

In managing behavioural change, it is necessary to understand what the drivers for change are. This can be for example, linking water and energy efficiency to help meet carbon targets and sustainability goals. The drivers for behavioural change in this case can include savings in water, energy, emissions, and costs; improving sustainability and environmental performance; and improving global or corporate social responsibilities. Behavioural change should also ideally be considered as a continuous cycle of improvement.

A key step in water-use efficiency is to measure and monitor consumption over time. The aim of this is to help understand exactly how much water is used, where, and why. This can be done using water consumption data, which is available from water billing (available in m^3) and/or by regular meter readings. The consumption data can be used to produce a water demand profile to help assess consumption over time and to benchmark water use against industry standard or other KPIs.

Benchmarking water use is straightforward for leisure centres that occupy whole buildings or premises [10]. Information needed for accurate benchmarking of these types of centres include at least a full year consumption data (m^3) and the use of actual, rather than estimated consumption data to ensure billing is based on meter readings and not an estimate of consumption. Where more data is available, for each year can be calculated to help understand consumption over time or to identify wastage or inform of necessary refurbishment.

Where a full year consumption data is not available, it is possible to use quarterly data to benchmark per day [10]. However, it is important to note that this may not be an accurate representation of water use given the likely seasonal fluctuation in footfall but can still be very useful for assessing and understanding how much water is being use, identify problem areas, and inform necessary interventions. Where footfall is transient and can significantly vary over the course of the year, it will be better to benchmark per floor area (m^2). For leisure centres that share premises or facilities, the best way to accurately assess water use will be to sub-meter the different areas of water use and water-use types.

A simple way to understand leisure centre water consumption and sewerage discharge is to develop a water balance model (Fig. 2). This is a numerical accounting of how water enters a site, how it is used, and how it is disposed of; and is done by mapping water flows through a site using information on water billing, sewerage and trade effluent discharge, meter readings, and site walk-around to ensure all water fittings and appliances are accounted.

A water demand profile is also a useful way of modelling consumption over time and can highlight changes in water use that may not be otherwise obvious. At least a whole year of data, and ideally a minimum of three years data, is needed and should be done quarterly, given the seasonal variation in footfall and water consumption in some leisure centres [12]. Average daily consumption can be calculated by dividing the number of opening days in the period by the annual consumption.

3.2 Wastewater heat recovery potential in leisure centres

Increasing the stock of low-carbon and/or renewable technology installations at the leisure centres like solar PV, biomass boilers, heat pumps, and WWHR will further reduce energy use and emissions and improve the carbon footprint and environmental performance of leisure centres with reasonable paybacks.

Although there is the potential for leisure centres to adopt heat recovery as a means of emissions and cost reductions, much remains to be done for the leisure sector to mobilise a critical mass for this to make meaningful impact for the sector. Even with the significant potential for energy savings through WWHR the capital cost for can be quiet substantial because most leisure centres operate year-round. Costs also vary widely because of the typically significant variation in types, sizes, ages, building fabric, etc., of leisure centres.

An example of the potential of WWHR in leisure centres was in the showers on the pool (wet) side of a community owned and operated leisure centre in rural Wales which has around 300 showers a week using four thermostatic mixed showers set at 40°C at an average flow rate of 12 litres per minute and average duration of 6 minutes. The water for the showers is heated by a pressurised boiler system set at 40°C which runs on oil gas. The plant where the

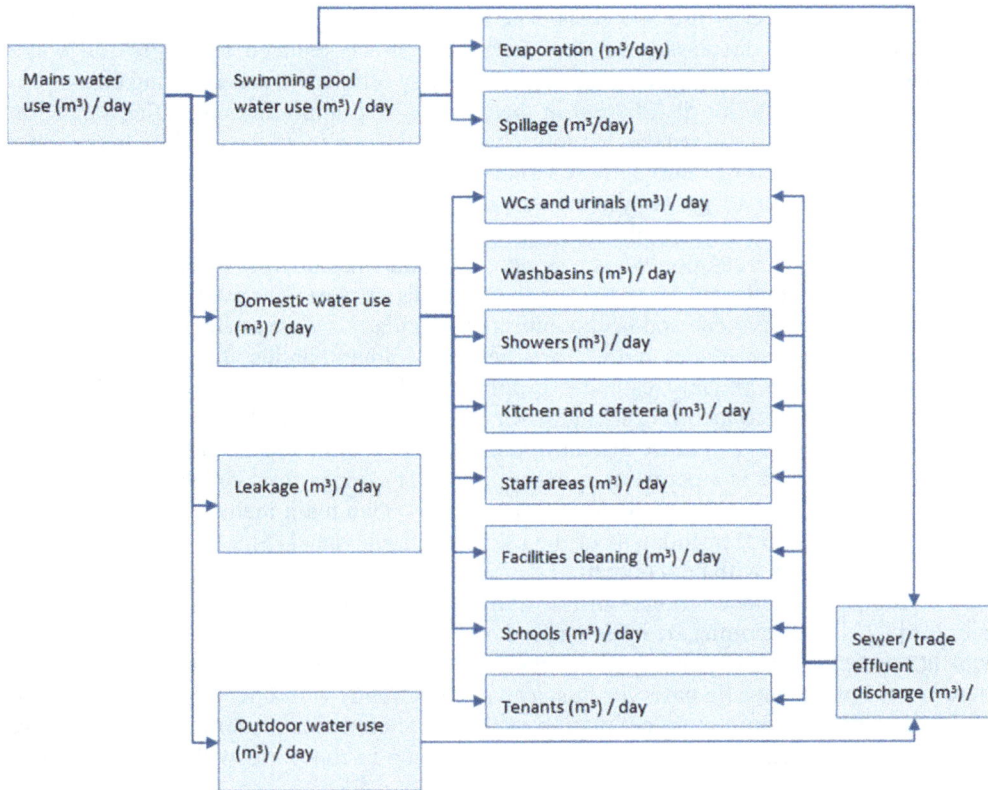

Figure 2: A water-balance model of a small rural leisure centre in Wales.

boiler is located is around 15 metres from the main plant room which is located on the dry side. Assuming total wastewater of 3,024 litres per day at 40°C and a flowrate of 45.6 litres per minute, a commercial wastewater heat recovery system can save 14,586 kWh energy, displace 4,000 $kgCO_2eq$ per year, save £1,245 per year with a payback of around 4 years with oil gas prices of £0.83 per m^3 and the price of electricity for pumping unit of £01.2 per kWh.

There is also the potential for other significant savings through other simpler, lower-cost interventions such as upgrading older fans and impellers to newer systems or the use of heat pumps. Opportunities for heat recovery are a lot larger when there is a pump type system than when it is only burning gas.

An ambitious intervention would involve the use of waste heat from chillers or condensers that would be otherwise dissipated for small and ambient heat networks within leisure centres. There is also the opportunity for localised networks to recover heat from different users though out the leisure centres, e.g., from gyms and other sport facilities to preheat pools, and vice versa in existing leisure centres, which is already becoming standard in a lot of new builds.

3.3 Improving systems control and optimisation

There are energy and cost savings potential from optimising of pool water circulation, air circulation, and even through water chemistry as the effectiveness of water treatment can have an impact on energy use and costs. The rate of removal of particulate material from pools can be greatly increased and the energy use much reduced by improving water circulation, thereby improving water quality and energy efficiency, and reducing costs.

Both circulation and air distribution in the pool hall are important mitigating factors in COVID-19 transmission, given the dramatic change in the quality of the air close to the water surface when moving for example, from a high-level extract to a low-level extract system. It is therefore important to get unidirectional airflow to reduce as many contaminants as possible, which is coincidentally the more energy efficient way of improving ventilation.

In terms of water circulation, we are developing guidelines on how much flow is needed and the minimum number of air turnovers in pool halls or water turnovers in the pool tank required to maintain hygiene and health and safety standards. More attention is also needed on water mixing characteristics and where water flows rather than just the outflow rate.

It is important to reliably estimate evaporation from swimming pools as incorrect calculations may lead to incorrect sizing of Air Handling Units (AHU) which could result in excessive humidity that can cause discomfort in the pool hall or even damage to building and equipment [15]. There is a potential in energy and cost savings from reliable estimation of evaporation from lower water and space heating costs. Two main methods can be used for this: analysis of physical phenomena or the use of empirical data [15].

Challenges remain with pool operation guideline orthodoxy. Many guidelines, such as the need to dump and replace 30 litres of hot water per bather [5] or guidelines on circulation rates [16]–[18], for example, are largely based on experience of good practice over the years, with little scientific underpinning. Yet, these are considered quite sacrosanct even though there is not much scientific basis for this. The general trends with operational guidelines in pools also tend to require more circulation, which inevitably means more energy use, more water dumping – both of which are in conflict of trying to run pools with minimal carbon footprint.

4 CONCLUSIONS

There is a significant potential for energy, emissions, and costs savings in leisure centres through water management, including water efficiency, use of low-carbon and renewable technology like WWRH, and through improved systems control and optimisation of water and air circulation, and through water treatment as water quality can be improved whist saving energy which can have the most impact if bundled together.

Many of these interventions are simple and available at zero or low-cost with no need to close centres and a payback within a budget year. The knowledge and the skill set to do this are available, but ambitious policy (e.g., Part L Building Regulations for commercial buildings) and financing instruments will be required to do this. In the UK, stepwise changes are moving the right direction, but it may be challenging to expect already stressed centres to buy into WWHR for example, even with rising fuel prices.

It is necessary to consider leisure centres to maximise the potential for moving towards Nearly Zero Energy Buildings (NZEB) because it is no longer sufficient to manage different areas and resources in isolation for any meaningful change in the way centres are operated and managed. Much work remains in terms of transition to electrification and decarbonisation of the grid, adequate insulation, use of heat pumps, and so on.

In a recent expert panel discussion on heat recovery potential of leisure centres we convened, the panellists agreed that it is not yet clear what the impact of the COVID-19 will be on future investment in green technology like WWHR, considering the long periods of closures and restrictions that have happened since March 2020. A lot of leisure centres in Wales are currently struggling financially as most leisure centres are LA or community owned and operate on a not-for-profit basis, so there may not to be a desire to invest until centres are back up and running and profitable again. Some leisure centres have also suffered infestation of filtration systems because chlorinated water has not been circulating during lockdown and there has been much damage to motors and electrical equipment due to being turned off for a while in cold, humid environments with a lot of plant needing to be replaced.

There is some evidence that government plan for net zero targets in the built environment and increasing austerity and cuts to public sector funding will push leisure centres towards increasingly ambitious emissions reductions to cut costs. The question remains as to whether policy will force the necessary move to electrification, particularly in rural areas that partially or fully operate on oil or gas.

ACKNOWLEDGEMENT

This research was undertaken within the Dŵr Uisce project, part funded by the European Regional Development Fund (ERDF) through the Ireland Wales Co-operation programme 2014–2020. We also acknowledge input from panellists of the Dŵr Uisce expert panel discussion on heat recovery potential of leisure centres: Lester Simmonds (Pool Sentry), Tony Gordon (Showersave), Adam Dyson (Buro Happold), and Michael Keane (Mark Eire BV).

REFERENCES

[1]   Howe, A., *Renewable Energy Potential for the Water Industry*, Environment Agency, 2009. ISBN: 978-1-84911-155-3.
[2]   Bello-Dambatta, A., Kapelan, Z. & Butler, D., Impact assessment of household demand saving technologies on system water and energy use. *British Journal of Environment and Climate Change*, **4**(2), pp. 243–260, 2014.
[3]   Passive House, Guidelines: Passive House concept for indoor swimming pools, 2021. https://passiv.de/downloads/05_guidelines_for_Passive_House_indoor_pools.pdf.
[4]   HM Government, Building Regulations, Approved Document L: Conservation of fuel and power, Volume 2: Buildings other than dwellings. © Crown Commonwealth, 2021.
[5]   Pool Water Treatment Advisory Group (PWTAG), Swimming pool water: Treatment and quality standards for pools and spas, 2017.
[6]   Resource Efficient Scotland, Resource efficiency in leisure centres and sports facilities guidance document, 2015. https://energy.zerowastescotland.org.uk/sites/default/files/100415%20RES%20Leisure%20Centres%20Sector%20Guide%20Final%20for%20Web.pdf.
[7]   Carbon Trust, Saving energy at leisure: Good practice guide GPG390, 2005.
[8]   European Parliament, EU Waste Framework Directive, Directive 2008/98/EC of the European Parliament and of the Council, 2008. https://eur-lex.europa.eu/legal-content/EN/ALL/?uri=CELEX%3A32008L0098.
[9]   Carbon Trust, Sector overview: Sport and leisure: Introducing energy saving opportunities for business CTV006, 2006.
[10]   Allen, A. et al., Guidance for new building projects, refurbishment, and facilities management: Procurement requirements for water efficiency. WRAP, 2010.
[11]   SAGE Environment and Modelling Group, Simple summary of ventilation actions to mitigate the risk of COVID-19, 2020.
[12]   Health and Safety Executive (HSE), Ventilation during the coronavirus (COVID-19) pandemic. © Crown Copyright, 2021. https://www.hse.gov.uk/coronavirus/working-safely/index.htm.

[13] Bello-Dambatta, A. et al., Guidelines for the evaluation and selection of sustainable water demand management interventions, Transitions to urban water systems for tomorrow report, 2014.

[14] Chartered Institution of Building Services Engineers (CIBSE), Energy use in sports and recreation buildings, 2001. https://www.cibse.org/getmedia/34def23a-c65b-405e-9dff-ce181c0b1e0d/ECG78-Energy-Use-in-Sports-and-Recreation-Buildings.pdf.aspx.

[15] Shah, M.M., Prediction of evaporation from occupied indoor swimming pools. *Energy and Buildings*, **35**(7), pp. 707–713, 2003.

[16] Pool Water Treatment Advisory Group (PWTAG), Swimming pool circulation pumps and variable speed drives (TN49), 2020. https://www.pwtag.org/pool-circulation-pumps-variable-speed-drives/.

[17] Chartered Institution of Building Services Engineers (CIBSE), Top tips: Ventilation in buildings, 2015. https://www.cibse.org/knowledge/knowledge-items/detail?id=a0q20000006oamlAAA.

[18] Health and Safety Executive (HSE), Health and safety in swimming pools. © Crown Copyright, 2018.

SECTION 3
ENERGY AND THE
BUILT ENVIRONMENT

CHOREOGRAPHING NETWORK-BASED INFLUENCES ON THE HOMEOWNERS' DECISION JOURNEY ABOUT ENERGY-RELATED RENOVATIONS

MARIA ISABEL ABREU, RUI A. F. DE OLIVEIRA & JORGE LOPES
Instituto Politécnico de Bragança, Portugal

ABSTRACT

Community engagement with the energy efficiency in housing is seen as a very challenging strategy that governments need to prioritize for climate agenda. Recent studies shed light on the underlying influences of some networks of actors in the homeowners' decision to improve the energy-efficiency of their homes. Understanding the accurate role that both social and professional networks within communities play among homeowners seems to be paramount for energy policies enlargement and effectiveness. In a network level perspective, interpersonal communication, which has so far been undervalued, appears to be an influential mode of trust and information to energy-related activities. An exploratory research has been carried out in order to gain a sense about the whole chain in action between the above mentioned actors as agents of change. Semi-structured in-depth interviews with different Portuguese stakeholders involved in the renovation journey of owner occupied single-family houses were conducted. The findings signpost that interpersonal networks play a trustworthy and beneficial source of information along the entire renovation decision-making journey. In general, these interpersonal connections are conveyed in the social network activating the homeowners' willingness to renovate, being a reliable channel for peer learning, guiding homeowners in the selection of professionals and acting also as an evaluative judgement tool to assess these same professionals in their technical advice. The results strengthen the view that the sense of neighbourliness, proximity and cooperation between citizens engaged in energy issues could pave the way to empower these utmost interpersonal networks which can have positive effects to encouraging house energy improvements.
Keywords: energy policies, homeowners, decision-making, social networks, renovation market professionals.

1 INTRODUCTION

Changing attitudes and behaviour remains a very necessary part of the efforts to reduce CO_2 emissions, however creating consumer demand for energy efficiency measures has been notoriously difficult to achieve. With buildings being a central part of our daily lives, the fact that 75% of the European building stock is still energy inefficient is a throwback for energy transition targets. In fact, around 74% of this stock belongs to the housing sector and is mostly inhabited by the owners, 70% on average in all European Union [1]. Furthermore, the weighted annual house energy renovation rate in the EU27 is still around 1% [1] with single-family houses being a challenge to policy-makers due to the many units, many owners and lack of shared organisations connecting them [2]. Additionally, people across major economies reduced visits to workplaces as teleworking become more normalised therefore time spent at home had increased by nearly 30% at the height of the lockdowns [3]. This time is being used to conduct activities that consume energy, leading to significant and complex shifts in homes energy demand [3]. Therefore, reducing green gas effect emissions in residential property owner-occupied is now more than ever a noteworthy challenge. The ultimate goal for energy policies is to make more non-interested householders attentive to home energy issues [4] since the official narratives are still much focus on already sensitized householders [5]. Moreover, there are cases where already interested citizens can abandon the renovation idea during the decision-making process before putting it into practice [6].

WIT Transactions on Ecology and the Environment, Vol 255, © 2022 WIT Press
www.witpress.com, ISSN 1743-3541 (on-line)
doi:10.2495/EPM220101

First and foremost, the success in renovation of the existing building stock to low carbon standards depends on a social, cultural and economic change as much as technical innovation [2], [7], [8]. Therefore, in a counterpart to the techno-economic model (which is a context-free approach), new theories take into account the influence of the social context around the homeowners where house renovations can be a topic of conversation and a matter of trust, friendship and shared values [9]. House renovation intentions may easily be overextended to encompass low carbon outcomes if an influencer or adviser involved is motivated to suggest such changes to the original arrangement [10]. Therefore, further than save energy and generate energy in houses, engaging the "wider community" in this crusade should be a determinant goal for policy-makers [11]. Hence, energy policies should target the networks of actors with which householders have close interaction with [12]. Thus, strengthening knowledge sharing and communitarian citizenship among neighbourhood residents through empowered networks of multiple intermediaries between government and homeowners is seen as a step forward to sustain innovative energy policies [13]. Intermediary actors from the social network and professionals from the ordinary and large repair-maintenance and improvement house market can be a means to governments interacting with homeowners further than via regulations and market manipulation [14]. Despite the potential impact that changes in the behaviour of these intermediaries seems to have, there appears to be slight policies that targets these groups [12] and lack of investigation on how they are influential to push homeowners to eventually purchase consecutive low-carbon renovation measures. Thus, this research was designed to facilitate the emergence of key issues about the significance that these types of networks have to lever the homeowners' interest in renovation in order to gain a sense of the whole chain in action where their belonging parties have a role to play. From the knowledge and perspective of Portuguese homeowners and professionals on the repair-maintenance and improvement house market some highlights are presented.

2 MATERIALS AND METHODS

A set of 21 semi-structured interviews were conducted under a qualitative in-depth exploratory study between October 2020 to September 2021 via video-conference and with a length of approximately 60 minutes. The interviews were conducted with eleven Portuguese homeowners of single-family buildings, three Portuguese energy advisors, four professionals from the craft business and three house energy systems certified installers. Studies suggest that seven interviews are sufficient for making relevant data emerge, however additional interviews were considered to provide redundancy and richness of data [15].

Studies identified some challenges when it comes to interviewing homeowners on their energy efficient decisions. Possible biases are related with: homeowners are highly involved in the decision and lack relevant technical and economical knowledge; they can demonstrate the commonly described "attitude action gap" (what consumers report as concerns or intentions has often little relation with what they do) and consumers tend to change their evaluation after high-involvement purchases [16]. Enquiring the energy advisors, craftspeople and installers was an efficient mean to get insights from the wide-ranging knowledge they gain through multiple in person consultations along their professional experience allowing detailed data to be collected with a limited number of interviews.

The participants were selected following purposeful sampling used in qualitative research for the identification and selection of information-rich cases related to a phenomenon (a specific criterion is used to select a particular sample where the aim is to collect in-depth information from the right respondents) [17]. The homeowners' sample (with ages between 39 and 65) varied according to actual possibilities and recommended by the professionals interviewed. Selecting homeowners that had already made some energy-related improvements or who were in the process of doing so was the leading criterion. More than one person per household to participate in the interview was much appreciated (six interviews included, in full time, the couples and in the other interviews, family members were present on some of the questions). The aim was to collect, as many opinions and experiences of householders as possible because the decision to renovate homes involves family negotiation and social interaction [5] with the woman playing an important role on this subject [18]. Six interviewees from the market side participants were also home renovators.

The interview's body were slightly adapted for each of the different interviewees' groups. Hypotheses raised from the theoretical background were used to define the fundamental interview topics which were used to prepare a group of direct (closed and open) questions. As an exploratory study, the open questions aimed to induce the interviewed to speak more about a particular subject to deepen the knowledge about it or even make new issues arise. Avoiding using some particular terms related with the study aim until the interviewees raised their own themes made part of the interview strategy. This enabled the respondents to describe their experiences in their own expressions and thereby to redefine the scope of the interview questions when pertinent. Thus, unplanned questions also emerged in response to the flow of homeowners' narratives to let the interviewees express themselves following the course of their thoughts and to guarantee that the interviews made a logical sense for them [17].

The full data assembled was confronted, analysed and discussed through a qualitative research method. The interviews were digitally recorded and then their transcripts were coded and categorized (using webQDA) to detect the most common themes in line with the research questions. A template of themes derived from the literature was developed previously from a conceptual model to facilitate examination [19]. All the interviewees were given pseudonyms from the transcription phase onwards. Inductive reasoning was used to make themes emerge from the raw data through repeated examination and comparison [19] and a realist theory approach to analyse how the previous knowledge about the subject was enlighten by the new data [20]. Qualitative findings from the small sample interviews followed a statistical theory of small sample qualitative research [15] however given the number of respondents, the results could not eventually be demonstrative of all cases and are limited to the universe of the circumstances found.

3 CONTEXTUAL BACKGROUND

3.1 About homeowner's decision for home renovation

Except in some urgent or out of ordinary situations, renovation is usually an act of free will and voluntary decision [16] involving a negotiation at the households' level [4], [16], [21], [22]. Step-by-step renovations, and not one-time events, dominate by far as a continuous improvement for living in and maintaining homes [23]. Energy renovations are merged with the other renovations [16], [18], [23] and step by step decision moments are usual to happen [9], [24]. For a long time, pay-back times, energy savings and rational approaches were considered as the main arguments for energy improvements. Nowadays, a social perspective brings to light that the underlying influences to begin with a renovation are shaped by homeowners' personal and social aspects [5], [9] which seems to boost the beginning of the decision process [5], [24] (Fig. 1). Therefore, it is not only a question of giving homeowners the right information but to see them as socially contextualized individuals who have domestic routines, carry embodied knowledge and skills and follow social norms of what is normal to do and to say [9]. It is inside this social environment where they get encouraged, seek and compare information and share experiences [4]. Decisions with high upfront payments, as house renovations, involve normally a voluntary external search [16] to form an opinion and seek reinsurance somewhere to supress feelings of uncertainty [12]. Simply becoming more receptive to information or otherwise by engaging in search behaviour [16].

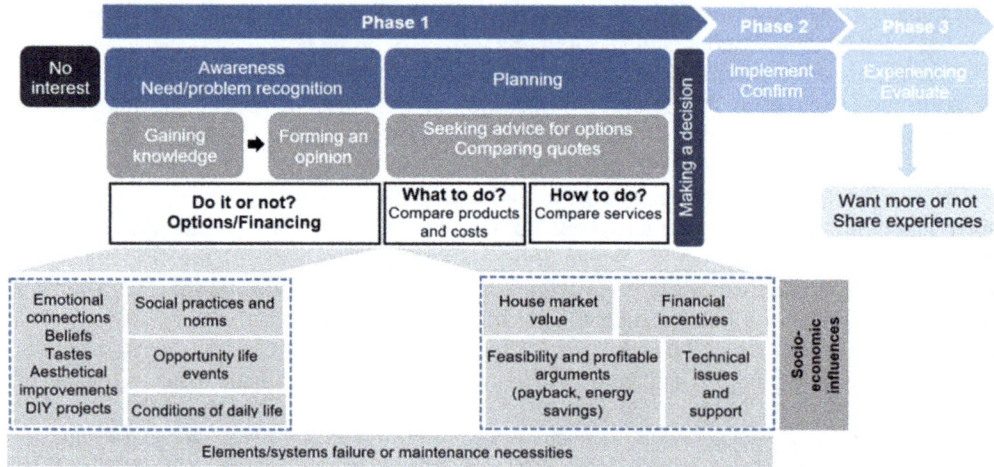

Figure 1: House renovation decision journey and influences [4], [5], [21], [22].

3.2 Knowledge networks around the homeowners for information and advice

Governments apparatus is characterised usually by not meeting homeowners face-to-face (bureaucratic reception on physical offices and websites, automated phone attendance, regulations and norms, etc.) [14]. To bond this distance and within governments vision for a society where empowered communities and voluntary action are widespread a comprehensive inclusion of non-state stakeholders from different levels in networks is needed [25].

It is at the social network level that interpersonal communication is a source of information [9] and where a specific mode of trust is built [12]. Face-to-face social relationships and the norms of reciprocity and trustworthiness formed within this networks leads inadvertently to members of a community identifying with each other, sharing values, ideas and favours [26]. Inside these networks strong ties between a close circle of family or friends are created, where there is a great deal of similarity [12]. However, weak ties also exist, a more fragile relationship due to occasional interactions, but in return provide access to new information, perspectives and experiences [12]. Voluntary peer learning process inside these networks can create and sustain know-how for householders' practices which are the ground for understanding what assemble house energy renovation actions [9]. Empowering these social linkages is argued as paramount for actual policies enlargement [11] in order to mobilize a "culture of energy" spread in the community [9].

Additionally, these social networks are also the field to get access and mobilize recourses with a specific purpose, for example when homeowners ask purposively for information using their interpersonal sources [26]. Homeowners search for trustable, simple, personally relevant and easily comparable information rather than exclusively technical and detailed [5]. Clear advice, reliable products and credible contractors are paramount [27]. Thus, the social networks are the field for informal conversations within socializing practices or for consultation of professionals from the social sphere around the homeowners [12]. Therefore, beyond the social network, where interpersonal connections are established, proximate networks of professional as agents of information and advice can work as mediating actors between governments and homeowners [13] ensuring other possible mode of trust to sustain house renovation decisions. Hence, despite an intense interaction between house owners and energy advisers has been sustained by policies, other potential actors seem to have a prominent role on this domain [4], [28], as for example craftspeople and installers. They seem to play an undervalued role in influencing both the adoption and use of domestic energy technology [10] and in shaping the ways in which energy-related renovations are carried out [9], [29]. As micro-enterprises they operate in informal networks of local skills and connections, often well-established depending on current local market demand [10]. Furthermore, their work is characterized by face-to-face interaction and by a near physical experience with the materiality of the house [14]. However, these professional's priorities and motivations not always align voluntarily with policy priorities [29]. Despite some intermediaries carry more less specialized knowledge in energy-related issues, as for example, certified energy advisers and specialized installers, craftspeople can act currently as non-certified energy consultants and for whom to go from single solution to system solutions can be a challenge [14]. The dispersed nature of these professionals' skills can make effective communication channels difficult to find [30] and the lack of access to competent or trustworthy contractors might hinder the final decision to renovate [31].

4 RESULTS AND DISCUSSION

4.1 Make homeowners interested: Gaining awareness, knowledge and forming an opinion

Firstly, in contrast to the expected governmental targets, this study disclosed that awareness raising is still far from being a common trigger for homeowners began to think about house renovation. The homeowners' willingness to renovate can be activated by informal daily conversations or by visible energy solutions neighbours and friends introduced in their houses. This can work as "eyes-catchers" for raising their awareness. However, from the stories collected, these circumstances need to be preceded by a pre-conceived disposition about making some improvements to the house. When asked about the cause of this preceding intentions the interviewees mainly reported issues related with personal tastes, physical condition of the house, conditions of domestic life and life stages. Attractiveness by indoor comfort, house design, aesthetics, DIY tasks and low carbon and smart innovative technologies (tech-savvy) for heating and cooling homes were the taste-related motivations identified. In some cases, house was no longer compatible with homeowners' personal and family living standards or homeowners were confronted with opportunity occasions some of them coincident with transitions periods in their life. The concern about social identity homeowners' house exhibit to the local community was also noticeable in the interviewees narratives which probably can be very related with the influence social values and norms inside the community have in shaping the homeowners' motivation [2], [9]. Thus, deep personal wishes and preferences already pre-existed in the majority of cases which seems to be activated into real actions with social network triggers (Fig. 2).

After some kind of interest on renovation was established, several homeowners expressed that they experienced to be overwhelmed by what they expected would be the next steps in

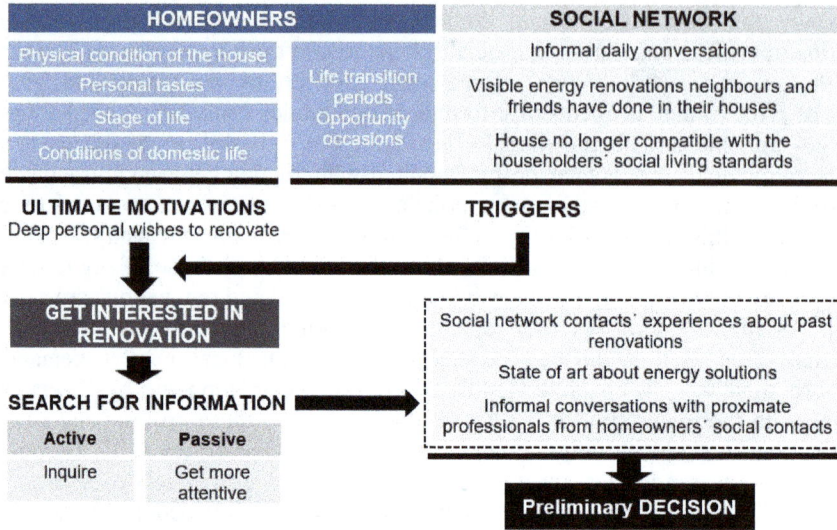

Figure 2: Ultimate motivations and the social network as a trigger to boost homeowners' renovation willingness.

the decision process. They sensed insecurity when they thought about seriously engage with energy savings measures due mainly to the uncertainty about financial capacity to support a potential renovation, fear of disruption of family life or effectiveness of improvements in indoor comfort. On this phase, homeowners feel the need for more information mainly to reinforce their preliminary idea about what they consider it would be the most suitable solutions to implement. Interpersonal contacts (from neighbourhood, local community and social media networks) were the sources from whom they receive early information, being passively more receptive to casual information or inquiring inside their close contacts. The raise of attention was, in great part, moulded by the peer learning process through what homeowners considered that reliable information is disseminated positively due to the fact that peers are "non-commercial" sources. Experiences on renovation and the state of art were the issues more pursued. However, professionals, normally belonging to householders' interpersonal contacts or recommended by someone they knew from their social interactions, are also informal sources for information. From the narratives analysis was clear these interpersonal channels can give encouragement and strengthen the idea of going forward with their intentions for renovation (Fig. 2). A particular finding is the fact that impersonal channels also were object of more attention during this phase, as media channels or campaigns.

4.2 The planning and decision phase: What and how to renovate

Once renovation willingness is established and an initial opinion is formed, homeowners started to search for more customized information to operationalize the renovation in more detail. Information collected exclusively from the social network could not unravel all the uncertainties about their particular circumstances and they seek tailor-made advice to minimize their concerns. A custom-made plan became a necessity and homeowners felt overloaded with the multifarious information from different sources. They qualified the experience as a difficult process to discern all the information, sometimes conflicting, that tend to make homeowners procrastinate in the decision. It was the case of Patrícia:

> "We were little bit confused about what we should invest first, if thermal insulation in the roof or install a biomass boiler and a solar panel. We tried to search for opinions from friends but at some point it was a little contradicting and advice were not always adapted to our circumstances".
>
> (Patrícia, 44, female)

Almost all the Portuguese homeowners interviewed stated they felt confused to whom contact to search for customized information. Again interpersonal mediators from the social network played a hinge role, inquired purposively or in opportunity moments by the homeowners. First, disclosing information about their experience on renovation solutions they encouraged, or discouraged, the homeowners about their intentions. Second, recommending professionals to guide about tailored options, help planning it and subsequently assist to materialize the renovation itself. Notably, the homeowners interviewed denoted that they have a scrambled and confused idea about the professional jurisdiction and skills of each of the market actors. They think all practitioners are at some point suited to give advice about energy issues for buildings. Despite some considered energy advisors as reliable professionals, respondents recognized that they did not knew their work and reputability so well as the craftspeople and installers work. Consequently, these last turned out to be privileged for contact. Established as small firms for a long time in the local market and doing several kinds of works in the buildings for which they gained a recognised reputation and trustiness. As was the case of Nuno, as demonstrated by the following quote:

> "I know their work. They already made some works in my house in the pass. It was clearly to me who I should contact in first place".
>
> (Nuno,44, male)

The professionals interviewed stressed that homeowners' reliability and trust in these proximate professionals was enhanced by their personal visits at homeowners' homes where skilled practitioners are able to see their non-verbal language which they think can provide a more reliable and realistic projection of energy savings for their particular case. Two of the energy advisors mentioned that they gather many times data on the house before any interactions with the householders. Assessing the property's attributes in terms of its location, looking for signs of life-stage or assessing the state of repair of the property were procedures revealed. Craftsmen and installers interviewed did not reported so much such practice. A possible result of the comprehensive training that certified energy advisors are subject to.

As mentioned before, homeowners have at this point a preliminary strong idea about what they wish to implement based mainly on their personal motivations and wishes. The majority of the homeowners recognized that when they contact a professional they bring with them preconceived ideas about what to do. Conflicts between what homeowners wished to implement and what was recommended by certified energy advisors was revealed in the narratives. The professional adviser seems to be often used by the homeowners to assess the properness of their already preferred solutions. This was pointed by one energy advisor:

> "Many times when we are asked to give support to homeowners they had already made some previous works suggested by a constructor or craftsman and it was not the most cost-effective".
>
> (Sofia, female, energy advisor)

Energy advisors had more difficulty to adapt to homeowners' requests when they believed homeowners' wishes were not the most efficient solutions to save energy and money. They seem to be more rigidly linked to regulations and technical framing and consequently less sensitized to homeowners' circumstances. Homeowners did not give positive feedbacks about this, they want to be valued and feel their opinions are important to the process. It is the case of Edgar, who assumed that was tempted to compare information with what their social network contacts considered about.

> "We had a little confront with the energy advisor, we wanted PV-panels and we try to convince him but he was not very receptive. We have friends with PV-panels on their homes and they are quite satisfied".
>
> (Edgar, 50, male)

One of the energy advisors interviewed, underlined that strong persuasive skills were needed to change some homeowners' ideas embedded in what he called "emotional choices". Other energy advisor highlighted that "was well prepared to offer technical advice but he lacked the training in communication skills to provide end-users with wider arguments".

On the other hand, the craftspeople and installers pointed several times they felt the necessity to filter what was technically feasible in order to suggest technologies which would meet the householder's preferences and simultaneously reduce energy costs. A reason found for they be preferably consulted is that homeowners detect and appreciate in this group an ability to "walk with" them through their preferences. They identified in them the capacity to select the appropriate solutions embracing the circumstantial information from the householders' context. Clients' motivation decrease when professional advice is too impersonal, as was revealed in the statement of Samuel:

> "I saw many times their expressions of disappointment when I suggested some solutions that they did not like or were in conflict with what they previously have in mind".
>
> (Samuel, craftsman)

However, the oral information received from an advisor in-person diminish the many fears and doubts, reinforce trust and is more effective than a simple simulation software tool available or a web platform as is revealed in the following quotation:

> "It is easier to trust experts as persons than in systems".
>
> (João, 49, male)

A particular aspect that came to the surface through the data analysis is that interviewees from craft business did not recognize themselves as building energy efficient practitioners. They accepted that they have been playing a role as energy advisers notwithstanding the majority never had an accurate training in a whole house renovation perspective which make sense since the majority of the energy renovations are made simultaneously with other renovations [23]. Despite the installers being the most trained, each of one is mostly expert on few specific technologies. This was revealed by Fábio:

> "The installer showed us a type of boiler to install but after a quick search in google we saw it was not the newest model. We ask one of our neighbours, who is a mechanical engineer, about his opinion. We became very disappointed

and distrustful about some installers since then… It is better to always confirm their technical recommendations".

(Fábio, 57, male)

Therefore, crafts business offered many times a contradictory advice in great part because house renovation market is fragmented between different types of professionals who are mainly concerned with business, to fit the economy of homeowners (and mainly with up-front costs) and to build a good reputation for future works [4], [10]. Thus, crafts business' information was seen by the respondents with more or less caution and distrust on some occasions. Homeowners interviewed reflected upon these professionals as still dull and poorly regulated, which affected their trust in their professional capacities because they consider they ultimately want to trade something motivated by profit. These uncertainties seem to be surpassed when they are recommended by interpersonal contacts or their reputation is blameless within the community.

Outcomes revealed also that when homeowners did not like or trust the installation service businesses or when the information provided by them about commissioning is unsatisfactory, they would be less likely to adopt a technology, which even is more reinforced when a still-developing technology is in discussion. Again homeowners tend to confirm with their interpersonal connections to seek their opinion. However, this was not the case for a technology adopter present in the sample, whose taste-related motivation make him run for high-tech solutions for his house. He was trialling a PV-panel despite the energy advisor did not recommend to be installed. One of the energy advisors, also stated from his experience, that the latest energy technology is seen as something that some people demand for a question of prestige and social status inside de community. Findings revealed as well that if the householder feels installation has been messier or disruptive than it should be it is likely they express dissatisfaction with the overall technology performance and can later on give a bad peer advice at the social network level. The experience of the installation itself and the support available after installation were mentioned as extremely important to rethink on doing more energy renovations in the future. Again homeowners either verified through their interpersonal networks if the whole experience they had was currently the way things are normal to happen about the installation itself and the efficiency of the technical solutions (Fig. 3).

Additionally, some respondents recognized that after they were fully committed to renovate, they became themselves opinion-leaders from the social networks and had already encouraged others to enrol in a network of collective learning.

Findings indicate that both craftspeople and installers can play a beneficial role more influential than it is expected by policy-makers and the governmental agenda and they tend to be preferable by homeowners compared with certified energy consultants. Despite professionals are more present in this phase of decision due to their expertise, social networks continue to be always a support that homeowners request both to search for guidance and to make cross-checking with the information or service given by professionals. Outcomes signpost that this is a search for positive feedback from whom they feel connected by proximity and by the circumstance they probably experience similar problems.

5 CONCLUSIONS

The results revealed that social networks matter in a home renovation decision and that interpersonal trust walks with the homeowners along the decision journey. As a field for network of collective learning and influence, social networks play a main role since the

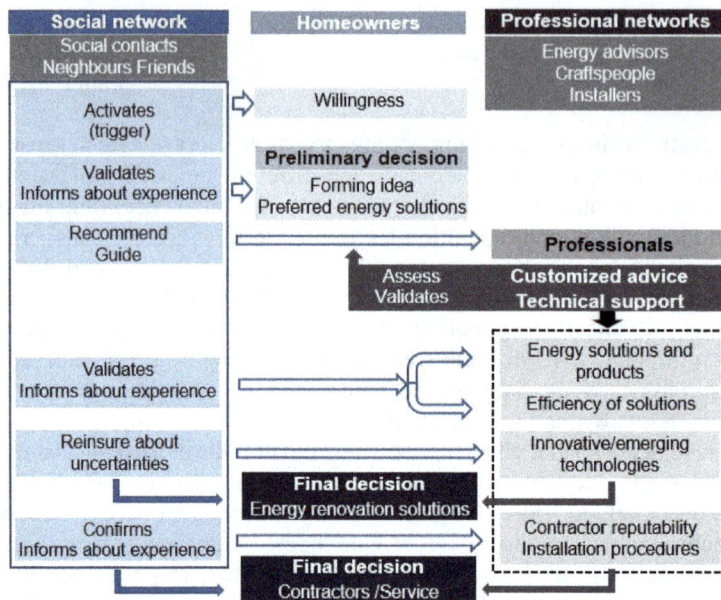

Figure 3:   Social and professional networks interaction in the homeowners' renovation decision-making process.

beginning of the homeowners' decision journey makes use of friendship and shared values. In the first place, the outcomes suggest that homeowners seem to be very attracted to display what is socially established and tend to replicate what others do. Undervalued established values and norms reproduced in the community through interpersonal connections through the "sense of community" can induce homeowners to go for home improvements. Hence, this can increase the interest of not yet committed energy users thus activating potential renovators. However, the findings also revealed that this social context influence is more likely to happen if homeowners have preceding intentions and preconceived ideas to renovate related with the physical condition of the house, the conditions of their domestic life, their personal tastes and stage of life. Despite professional support being crucial to definitely engage homeowners due to their expertise to make an accurate technical selection of products and services, knowledge and experience available through homeowners' social network, mediated through their peers, acts as a guide to recommend professionals and as a cross checking tool about their advice and reputability. In sum, these interpersonal networks that homeowners inquire purposively to validate the information are used to audit professional advice and encompass the entire decision-making journey.

The outcomes also signpost the key role of trustworthy networks of small firms of craftspeople and installers who act as energy consultants surpassing many times certified energy agents. Their positive reputation inside the homeowners' interpersonal network still make them as favourites to ask for advice despite these craft businesses and installation services work under a fragmented market of practitioners. The absence of knowledge in a whole perspective about how to make homes more energy efficient can often result in a contradictory technical advice that affects homeowners' trust in their capacities. Many times concerned with their business, they are unlikely to orient their behaviour according to policies of environmental and energy values. However, the study points that homeowners appreciate their nontechnical skills on an adaptive capacity to assimilate contextual and circumstantial information from the householders and to think in line with their wishes and needs. In fact, homeowners tend to seek validation about previous or preliminary choices, not always being very responsive to energy efficiency rational arguments. This is something energy advisors are not so flexible to act upon.

In sum, interpersonal and established professional networks interrelate and complement each other acting the second as a technical support as expected and the first as a guide, cross checking and reinforcement tool for homeowners bringing to the light the importance of trusted messengers to support homeowners' decisions. A continuous support is essential to ensure that homeowners acquire the provision they need to progress in the decision for which technical information is an essential feature but trust and communication skills do not get behind.

This study gave rise to a number of questions for future research and policies. In the short–medium run, policies need to provide the positive social context to established a strong local community of knowledge networks to target different types of homeowners' profiles and lift their interest in home renovation. On the other hand, changing the institutional arrangement and leaving it, in part, for more traditional actors is necessary. It is the case for letting craftspeople and installers also take part as trust intermediaries to be deployed at specific stages of the decision-making process. A closer cooperation between them and the certified energy advisors seems to be required, where the first ones need to acquire more comprehensive technical capabilities and the seconds need to take lessons from the well-established network market of craft businesses. Challenging means towards the energetic transition in homes can be in the ability of governments to involve strategic intermediaries from different levels promoting dialogue and partnerships. Finally, the study is developed based on empirical evidence from a qualitative study on a specific region and with a limited sample size of renovation adopters. Extending the scope of data collection can probably generate further findings. Further studies could also test a wider set of other factors related with the social network situation, such as specific cultural community aspects and neighbourhood socio-economic characteristics. Also a research among homeowners is needed to provide more insights about which of homeowners' profiles are motivated by what type of energy service design in a way that factors that are foremost important for each stage of decision could be investigated.

REFERENCES

[1] European Commission, The Renovation Wave: The European Green Deal. https://ec.europa.eu/energy/topics/energy-efficiency/energy-efficient-buildings/renovation-wave_en#a-renovation-wave-for-europe. Accessed on: 6 Oct. 2021.

[2] Gram-Hanssen, K., Ole Jensen, J. & Friis, F., Local strategies to promote energy retrofitting of single-family houses. *Energy Efficiency,* **11**, pp. 1955–1970, 2018.

[3] International Energy Agency, Energy Efficiency 2020, 2020. https://iea.blob.core.windows.net/assets/59268647-0b70-4e7b-9f78-269e5ee93f26/Energy_Efficiency_2020.pdf. Accessed on: 10 Oct. 2021.

[4] Broers, W., Vasseur, V., Kemp, R., Abujidi, N. & Vroon, Z. Decided or divided? An empirical analysis of the decision-making process of Dutch homeowners for energy renovation measures. *Energy Research and Social Science*, **58**, 101284, 2019.

[5] Wilson, C., Crane, L. & Chryssochoidis, G., Why do homeowners renovate energy efficiently? Contrasting perspectives and implications for policy. *Energy Research and Social Science*, **7**, pp. 12–22, 2015.

[6]   Boza-Kiss, B., Bertoldi, P., Della Valle, N. & Economidou, M., One-stop shops for residential building energy renovation in the EU. Report number: JRC125380, European Commission, 2021.

[7]   Nilsson, M., Zamparutti, T., Petersen, J.E., Nykvist, B., Rudberg, P. & McGuinn, J., Understanding policy coherence: Analytical framework and examples of sector–environment policy interactions in the EU. *Environmental Policy and Governance*, **22**, pp. 395–423, 2012.

[8]   Ravetz, J., State of the stock: What do we know about existing buildings and their future prospects? *Energy Policy*, **36**, pp. 4462–4470, 2008.

[9]   Bartiaux, F., Gram-Hanssen, K., Fonseca, P., Ozolina, L. & Christensen, T.H., A practice–theory approach to homeowner' energy retrofits in four European areas. *Building Research and Information*, **42**(4), pp. 525–538, 2014.

[10]   Owen, A. & Mitchell, G., Unseen influence: The role of low carbon retrofit advisers and installers in the adoption and use of domestic energy technology. *Energy Policy*, **73**, pp. 169–179, 2014.

[11]   Platt, R., Cook, W. & Pendleton, A., *Green Streets, Strong Communities*, IPPR: London, 2011.

[12]   Wilde, M. de, Designing trust: How strategic intermediaries choreograph homeowners' low-carbon retrofit experience. *Building Research and Information*, **47**, 2018.

[13]   Wilde, M. de, The sustainable housing question: On the role of interpersonal, impersonal and professional trust in low-carbon retrofit decisions by homeowners. *Energy Research and Social Science*, **51**(4), pp. 138–147, 2019.

[14]   Fyhn, H., Søraa, R.A. & Solli, J., Why energy retrofitting in private dwellings is difficult in Norway: Coordinating the framing practices of government, craftspeople and homeowners. *Energy Research and Social Science*, **49**, pp. 34–142, 2019.

[15]   Galvin, R., How many interviews are enough? Do qualitative interviews in building energy consumption research produce reliable knowledge? *Journal of Building Engineering*, **1**, pp. 2–12, 2015.

[16]   Baginski, J.P. & Weber, C., A consumer decision-making process? Unfolding energy efficiency decisions of German owner-occupiers. HEMF Working Paper, No 08/2017, 2017.

[17]   Quinn Patton, M., *Qualitative Research and Evaluation Methods: Integrating Theory and Practice*, SAGE: Thousand Oaks, CA, 2015.

[18]   Sunikka-Blank, M., Galvin, R. & Behar, C., Harnessing social class, taste and gender for more effective policies. *Building Research and Information*, **46**(1), pp. 114–126, 2018.

[19]   King, N., Using templates in the thematic analysis of text. *Essential Guide to Qualitative Methods in Organizational Research*, eds C. Cassell & G. Symon, SAGE: London, pp. 256–270, 2004.

[20]   Crouch, M. & McKenzie, H., The logic of small samples in interview-based qualitative research. *Social Science Information*, **45**, pp. 483–499, 2006.

[21]   Klöckner, C.A. & Nayum, A., Specific barriers and drivers in different stages of decision-making about energy efficiency upgrades in private homes. *Frontiers of Psychology*, **7**, p. 1362, 2016.

[22]   Ebrahimigharehbaghi, S., Qian, Q.K., Frits, M. & Visscher, H.J., Unravelling Dutch homeowners' behaviour towards energy efficiency renovations: What drives and hinders their decision-making? *Energy Policy*, **129**, pp. 546–561, 2019.

[23] European Commission, Comprehensive study of building energy renovation activities and the uptake of nearly zero-energy buildings in the EU, Final Report, 2019. https://ec.europa.eu/energy/studies/comprehensive-study-building-energy-renovation-activities-and-uptake-nearly-zero-energy_en. Accessed on: 30 Nov. 2021.

[24] Judson, E.P. & Maller, C., Housing renovations and energy efficiency: insights from homeowners' practices. *Building Research and Information*, **42**(4), pp. 501–511, 2014.

[25] Van der Heijden, J., The new governance for low-carbon buildings: mapping, exploring, interrogating. *Building Research and Information*, **44**, pp. 575–584, 2016.

[26] Lin, N., Building a network theory of social capital. *Connections*, **22**(1), pp. 28–51, 1999.

[27] Galvin, R. & Sunikka-Blank, M., The UK homeowner-retrofitter as an innovator in a socio-technical system. *Energy Policy*, **74**, pp. 655–662, 2014.

[28] Nair, G., Gustavsson, J. & Mahapatra, K., Factors influencing energy efficiency investments in existing Swedish residential buildings. *Energy Policy,* **38**(6), pp. 2956–2963, 2010.

[29] Risholt, B. & Berker, T., Success for energy efficient renovation of dwellings: Learning from private homeowners. *Energy Policy,* **61**, pp. 1022–1103, 2013.

[30] Galvin, R. & Sunikka-Blank, M., The UK homeowner retrofitter as an innovator in a socio-technical system. *Energy Policy*, **74**, pp. 655–662, 2014.

[31] Weiss, J., Dunkelberg, E. & Vogelpohl, T., Improving policy instruments to better tap into homeowner refurbishment potential: Lessons learned from a case study in Germany. *Energy Policy,* **44**, pp. 406–415, 2012.

MACHINE LEARNING APPROACH FOR PREDICTIVE MAINTENANCE IN AN ADVANCED BUILDING MANAGEMENT SYSTEM

SOFIA AGOSTINELLI[1] & FABRIZIO CUMO[2]
[1]Department of Astronautics, Electrical and Energy Engineering, Sapienza University of Rome, Italy
[2]Department of Planning, Design, and Technology of Architecture, Sapienza University of Rome, Italy

ABSTRACT

Predictive maintenance is a concept linked to Industry 4.0, the fourth industrial revolution that monitors equipment's performance and condition during regular operation to reduce failure rates. The present paper deals with a predictive maintenance strategy to reduce mechanical and electrical plant's malfunctioning for residential technical plant systems. The developed strategy can guarantee a tailored maintenance service based on machine learning systems, drastically reducing breakdowns after a maximum period of 3 years. The developed strategy evaluates an acceptable components failure rate based on statistical data and combines the average labour costs with the duration of each maintenance operation. The predictive strategies are elaborated on the minimum cost increase necessary to achieve the abovementioned objectives. A case study based on a 3-year-period has been conducted on a modern residential district in Rome composed of 16 buildings and 911 apartments. In particular, the analysis has been performed considering mechanical, electrical and lighting systems supplying the external and common areas, excluding the apartments, to avoid data perturbation due to differential user's behaviours. The overall benefits of predictive maintenance management through Big Data analysis have proven to be the substantial improvements in the overall operation of different plants as mechanical and electrical plants of residential systems.

Keywords: BIM environment, facility management, predictive maintenance, security management, energy management, digital twin.

1 INTRODUCTION

Due to the widespread use of industrial machinery, a huge amount of data can be collected every day, resulting in increased attention of production managers and data analysts to potential applications. Predictive maintenance (PdM) is a prominent approach in this regard, which intends to monitor and analyze a system in real-time to identify and prevent potential maintenance needs promptly [1]. Furthermore, PdM intends to prevent incipient failures by paying attention to unusual behaviours providing a just-in-time maintenance intervention; such interventions are necessary to ensure both the quality and safety of devices and prevent unnecessary costs [2]. Building maintenance is a crucial section of facility management (FM), mainly because of maintenance-related expenditures for at least 65% of FM costs each year [3].

Several studies have investigated PdM models. For instance, Quatrini et al. [4] performed an exhaustive literature review in this field. They mentioned four main areas: (a) PdM fundamentals and implementation; (b) PdM strategies; (c) inspections and replacement plans; and (d) prognosis. Many studies focused on PdM plans reported Remaining Useful Life (RUL) as a condition index. In addition, Zhou et al. [5] reported a reliability-oriented model based on a PdM continuous monitoring system, which is based on the hypothesis that such systems are prone to continuous degradation that can be monitored. Such degradation processes can be modelled using the Markov decision process [6], proportional hazards model [7], a gamma process [8], or Monte Carlo simulation [9], [10]. Berka and Macek [11] presented a model for fault identification to perform adequate maintenance interventions.

WIT Transactions on Ecology and the Environment, Vol 255, © 2022 WIT Press
www.witpress.com, ISSN 1743-3541 (on-line)
doi:10.2495/EPM220111

Susto et al. [12] reported a PdM system based on the availability of the current values of the physical factors acting on the production process and Support Vector Machines (SVMs). In order to detect faulty and non-faulty states of the machines, SVMs are applied, which can estimate the distance from failure as the equipment RUL. Wang et al. [13] developed a prognostic model that can link the extended Kalman filter with a first-order perturbation technique. They developed a cost-based predictive maintenance (CDPM) framework and reported their model as a case study. Wang et al. developed a cloud-based paradigm to predict maintenance using a mobile agent to allow in-time access to information and to apply them to enhance the accuracy and reliability of the fault identification process, estimating service life, and schedule maintenance procedures [14]. Nevertheless, their model does not contain a valid prediction algorithm for condition prediction. Ren and Zhao introduced a framework based on the Internet of Things (IoT) used to manufacturing, and Operation and Maintenance (O&M) data obtaining a procedure for a decision-making technique to predict the need for maintenance, including decision tree, k-Means, support vector machine (SVM), and neural network [15]. Nevertheless, their study does not explain how to apply this technique and the prediction procedure. Finally, Cheng et al. performed a study to compare marker-based Augmented Reality (AR) and marker-less AR for indoor maintenance and decoration [16].

Nevertheless, the latter model is solely about maintenance and operational information and does not contain maintenance planning. Kang and Hong developed software intended to incorporate building information modelling (BIM) effectively with a GIS-based facilities management (FM) system [17]. In addition, they reported a BIM/GIS-based prototype intended to extract, transform, and load data from BIM and GIS to integrate FM data. Ayvaz and Alpay developed a maintenance system based on available data to predict production lines using machine learning techniques and IoT [18]. Vafaei et al. presented an alarm system based on fuzzy logic to predict early equipment degradation in a car manufacturing line, emphasising minimizing costs related to sudden failures [19]. Wei et al. introduced a conditional maintenance framework intended to identify the optimal action to reduce the mean cost rate as much as possible [20]. Cheng et al. introduced a data-based model to predict maintenance of mechanical, electrical, and plumbing components based on information modelling and IoT [21].

Based on what was mentioned above, the research gap contains (1) inadequate data integration for predictive maintenance; (2) absence of good predictive patterns; and (3) no description of predictive procedures. Therefore, this paper's main novelty and contribution is to provide an intelligent predictive maintenance strategy for mechanical, electrical, and plumbing sections of building facilities, like HVAC systems, electrical parts, and lighting, that have a crucial role in ensuring the functionality of buildings.

2 CASE STUDY

The Rione Rinascimento complex hosts 3,000 tenants in 950 apartments divided into seven building complexes. Each complex consists of three eight-floor buildings. Therefore, it can be defined as an eco-neighbourhood. It is powered by 65% of renewable sources and 100% only regarding the thermal consumption for the production of hot water and air conditioning (Fig. 1).

This result has been achieved thanks to the building envelope's high efficiency and installing the largest European geothermal power plant for residential use working through 190 geothermal probes of 150 m depth, allowing favourable heat exchange with the ground in winter and summer conditions. In addition, a bio-mass cogeneration plant coupled with the geothermal plant allows reaching economic savings of 40% and a decrease in CO_2 emissions of 50%.

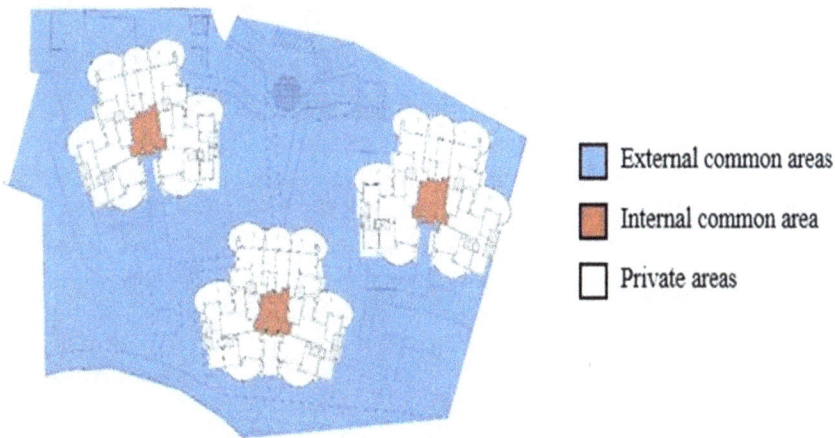

Figure 1: Building module in Rione Rinascimento complex.

The case study is significant due to peculiar conditions related to energy and safety management and building maintenance strategies. In fact, from the energy management point of view, it is considered as self-sufficient; considering the safety management, it is characterized as a well-defined system, physically separated from the surrounding urban context; from the maintenance point of view, it allows obtaining numerous and structured data as it is based on a global services contract. This paper investigates the maintenance aspect with a view to the optimization of planned maintenance processes through predictive systems.

2.1 Maintenance system

The maintenance datasheets are the starting point of the entire process, as they provide the necessary information about the analysis of the considered elements divided into categories/disciplines. In the case study, customized datasheets have been created for the mechanical and electrical systems. However, both are only considered in the internal and external common areas, thus excluding spaces for private use.

The used maintenance datasheets only consider activities and elements that can cause faults and malfunctions involving replacing single components. These sheets are regularly compiled for each scheduled maintenance operation.

3 PROPOSED PREDICTIVE MAINTENANCE STRATEGY

The PdM contains three key stages [22], including data acquisition: obtaining data from different sources is equally essential for PdM. As is structured, the data acquisition process is fully automated. The second one is unstructured, and the collection can be partially entrusted to the operator. Hence, accurate validation is required. Therefore, data processing should be performed because this process is crucial for dataset cleaning and to analyze data to assess its consistency with the physical phenomenon. The last process is the maintenance decision-making stage. It contains two main groups: (a) fault identification and (b) fault prediction. In this study, a maintenance strategy is presented for mechanical, electrical and lighting systems supplying the external and common areas excluding the apartments to avoid data perturbation due to differenced user's behaviours. The proposed strategy implementing are as follows:

A. Collecting data;
B. Determining the parameters affecting the system outputs;
C. Determine the optimal values of the obtained parameter;
D. Fault identification model;
E. Fault prediction intelligent strategy for electrical and mechanical systems.

In order to optimize the timing of the maintenance operations, the study started by considering the maximum initial values of the maintenance intervals taken from the Italian and European technical standards. A threshold value of tolerability of faults detected at the maintenance visit was chosen, estimating around 5% for all types of implemented components. This choice is related to factors such as avoiding excessively high malfunctioning values that would compromise the functionality of the systems or at least penalize the quality of service, preventing an excessive increase in costs associated with scheduled maintenance activities.

The proposed methodology is based on a statistical evaluation of the deviation between the failure rate of a single device and the set threshold value (as mentioned above, 5% was chosen for uniformity and simplicity in this case study). If this value is lower than the threshold, subsequent maintenance will be scheduled according to the time interval already provided by the standard; on the other hand, if the value of the failure rate is higher than the threshold limit, the scheduled time interval will be shortened, and subsequent maintenance will be anticipated. The shortened period will follow the percentage value exceeding the 5% threshold according to eqn (1):

$$Mp = Mppl - [Mppl \times (fr - frth)] \qquad (1)$$

where fr is the real failure rate, frth is the threshold limit of the failure rate, Mppl is the planned maintenance period, Mp is the shortened maintenance period.

Whenever a scheduled maintenance cycle of components is completed, the failure rate statistics must be evaluated and the timing for the following maintenance schedule is updated. When the failure rate curve falls below 5%, this means that the chosen interval is adequate and does not need to be shortened further. However, supposing robust requirements for continuity of service of the installations are needed, such as for some essential electrical power supply system components, it will be sufficient in that case, to lower the threshold limit to reach the desired reliability values, varying the time interval of the maintenance operations accordingly.

4 RESULTS AND DISCUSSIONS

The proposed building management system can achieve much more considerable efficiency by configuring the Digital Twin of the building through the integration of the BIM model aimed at AIM (asset information model) [23] with information systems, BMS, IoT, machine learning, mixed reality for the optimization and automation of maintenance activities.

In addition, working on the BIM model as the core of the management system architecture, it is possible to improve the potential related to physical/spatial information for space management, optimizing maintenance activities, integrating machine learning systems and rule-based methods such as association rule mining [24]. Therefore, it is helpful to create a hierarchy in the classification of spaces, using machine learning techniques defined as "clustering", automatically identifying groups/classes of similar spaces for their digital representation.

The proposed system, therefore, involves the integration of three components: (i) data related to BIM Model objects (Autodesk Revit); (ii) data flow programmed through visual programming systems (Autodesk Dynamo) relating the AI systems with the BIM model bidirectionally; (iii) Artificial Intelligence (AI) algorithm through Python language improving efficiency and compatibility with flows programmed through Dynamo [25]. Furthermore, the AI system can use mixed techniques.

As a preliminary result, in order to test the methodology, the following components with their base maintenance intervals, obtained from the related regulations were taken into account in the period 2018–2020. Those components are the only ones with complete available data for a suitable period (at least 3 years of continuous data) (Table 1).

Table 1: Base maintenance frequency plan.

HVAC	Frequency	Electrical	Frequency
Boiler	Annual	Electrical panel	Monthly
Piping insulation	Annual		
Cooling system	Annual	Lightning systems lamps	Quarterly
AHU belt and motion parts	Monthly		
AHU flat filters	Quarterly	Lightning system equipment	Monthly
Valves	Monthly		

By graphically describing the 3-year data obtained following the maintenance datasheets as described in paragraph 2, it is possible to evaluate the trends in the failure rate (shown on the y-axis) as a function of the shortening of the maintenance intervals carried out according to eqn (1).

Fig. 2 shows how the failure rate tends to fall below 5% after a period varying according to the performed maintenance cycles. The threshold is reached after less than a year for monthly maintenance activities while quarterly maintenance activities take about two and a half years; however, if the number of maintenance cycles is assessed (12 for the electrical panel and nine for air handling unit (AHU) filters, respectively), it can be observed that the trend is similar. Furthermore, the percentages of shortening maintenance intervals are also similar and range from 24% for the electrical panel and 17% for AHU.

Therefore, it is always possible to reach the desired failure rate level with a proportional increase in expense as maintenance intervals are shortened. Therefore, to optimize the operational cost–benefit, it is crucial to precisely identify the value of the most suitable threshold limit for the specific conditions.

5 CONCLUSIONS

Big Data is becoming a preponderant issue in many technological fields; predictive maintenance should develop innovative methodologies to increase a technical plant's reliability levels towards a near-zero-failure rate. In the present paper, initial indications, even coming from partial data, show the possibility strategies to obtain the most suitable threshold failure rate values for each kind of equipment. With the help of business intelligence software, it will also be possible to organise data from different sources and carry out forecast simulations even in the absence of specific data. Such an approach is confirmed by the failure of traditional supervised analyses, depending on great difficulties in relating involved variables to their actual impact on the maintenance strategies. So, the collection of a

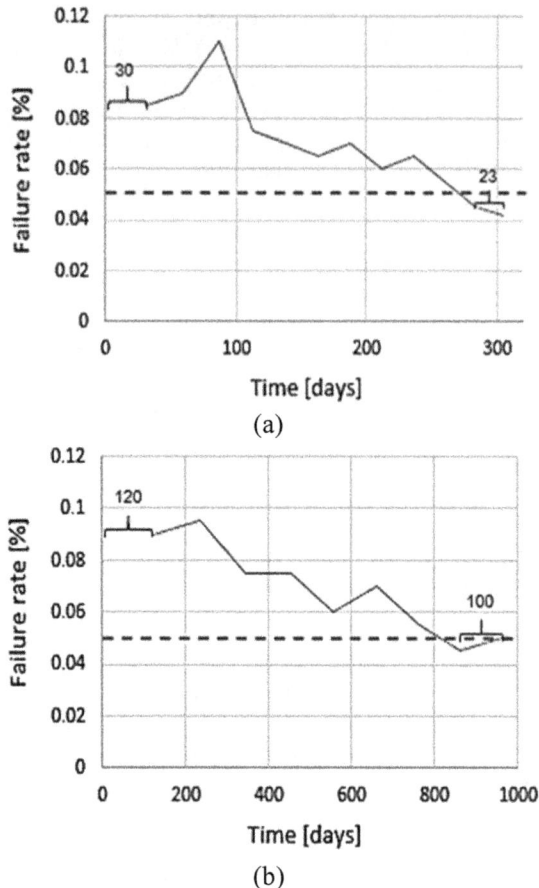

Figure 2: (a) Electrical panel monthly maintenance, and failure rate reduction; and (b) Air handling unit quarterly maintenance and failure rate reduction.

significant amount of maintenance data from different building systems, coupled with a data acquisition tool able to filter inadequate information, will improve the predictive maintenance strategy in public and private organizations. The preliminary results highlighted this in the present paper.

REFERENCES

[1] Selcuk, S., Predictive maintenance, its implementation and latest trends. *Proceeding of the Institution of Mechanical Engineers Part B, Journal of Engineering Manufacturing*, **231**(9), pp. 1670–1679, 2017.

[2] Krishnamurthy, L. et al., Design and deployment of industrial sensor networks: Experiences from a semiconductor plant and the North Sea. *Proceedings of the 3rd International Conference on Embedded Networked Sensor Systems*, pp. 64–75, 2005.

[3] Eastman, C., Teicholz, P., Sacks, R. & Liston, K., *BIM Handbook: A Guide to Building Information Modeling for Owners, Managers, Designers, Engineers and Contractors*, John Wiley and Sons, 2011.

[4]   Quatrini, E., Costantino, F., Di Gravio, G. & Patriarca, R., Condition-based maintenance: An extensive literature review. *Machines*, **8**(2), p. 31, 2020.

[5]   Zhou, X., Xi, L. & Lee, J., Reliability-centered predictive maintenance scheduling for a continuously monitored system subject to degradation. *Reliability Engineering and Systems Safety*, **92**, pp. 530–534, 2007. DOI: 10.1016/j.ress.2006.01.006.

[6]   Sheu, S., Chang, C., Chen, Y. and George, Z., Optimal preventive maintenance and repair policies for multi-state systems. *Reliability Engineering and Systems Safety*, **140**, pp. 78–87, 2015. DOI: 10.1016/j.ress.2015.03.029.

[7]   Tian, Z. & Liao, H., Condition based maintenance optimization for multi-component systems using proportional hazards model. *Reliability Engineering and Systems Safety*, **96**(5), pp. 581–589, 2011. DOI: 10.1016/j.ress.2010.12.023.

[8]   Liao, H., Elsayed, E.A. & Chan, L., Maintenance of continuously monitored degrading systems. *European Journal of Operational Research*, **175**, pp. 821–835, 2006. DOI: 10.1016/j.ejor.2005.05.017.

[9]   Marseguerra, M., Zio, E. & Podofillini, L., Condition-based maintenance optimization by means of genetic algorithms and Monte Carlo simulation. *Reliability Engineering and Systems Safety*, **77**, pp. 151–165, 2002.

[10] Barata, J., Soares, C.G., Marseguerra, M. & Zio, E., Simulation modelling of repairable multi-component deteriorating systems for 'on condition' maintenance optimisation. *Reliability Engineering and Systems Safety*, **76**, pp. 255–264, 2002.

[11] Berka, J. & Macek, K., Effective maintenance of stochastic systems via dynamic programming. *Proceedings of 19th Technical Computing Prague Conference*, p. 346, 2011.

[12] Susto, G.A., Schirru, A., Pampuri, S., Mcloone, S., Member, S. & Beghi, A., Machine learning for predictive maintenance: A multiple classifier approach. *IEEE Transactions on Industrial Informatics*, **11**(3), pp. 812–820, 2015.

[13] Wang, Y., Gogu, C., Binaud, N., Bes, C., Haftka, R.T. & Kim, N.H., A cost driven predictive maintenance policy for structural airframe maintenance. *Chinese Journal of Aeronautics*, **30**(3), pp. 1242–1257, 2017. DOI: 10.1016/j.cja.2017.02.005.

[14] Wang, J., Zhang, L., Duan, L. & Gao, R.X., A new paradigm of cloud-based predictive maintenance for intelligent manufacturing. *Journal of Intelligent Manufacturing*, **28**(5), pp. 1125–1137, 2017.

[15] Ren, S. & Zhao, X., A predictive maintenance method for products based on big data analysis. *Proceedings of the International Conference on Materials Engineering and Information Technology Applications (MEITA)*, pp. 385–390, 2015.

[16] Cheng, J.C.P., Chen, K. & Chen, W., Comparison of marker-based AR and marker-less AR: A case study on indoor decoration system. *Proceedings of Lean and Computing in Construction Congress (LC³)*, Heraklion, Greece, pp. 483–490, 2017.

[17] Kang, T.W. & Hong, C.H., A study on software architecture for effective BIM/GIS-based facility management data integration. *Automation in Construction*, **54**, pp. 25–38, 2015. DOI: 10.1016/j.autcon.2015.03.019.

[18] Ayvaz. S. & Alpay, K., Predictive maintenance system for production lines in manufacturing: A machine learning approach using IoT data in real-time. *Expert Systems and Applications*, **173**, 114598, 2021. DOI: 10.1016/j.eswa.2021.114598.

[19] Vafaei, N., Ribeiro, R.A. & Camarinha-Matos, L.M., Fuzzy early warning systems for condition based maintenance. *Computers and Industrial Engineering*, **128**, pp. 736–746, 2019. DOI: 10.1016/j.cie.2018.12.056.

[20] Wei, G., Zhao, X., He, S. & He, Z. Reliability modeling with condition-based maintenance for binary-state deteriorating systems considering zoned shock effects. *Computers and Industrial Engineering*, **130**, pp. 282–297, 2019. DOI: 10.1016/j.cie.2019.02.034.

[21] Cheng, J.C.P., Chen, W., Chen, K. & Wang, Q., Data-driven predictive maintenance planning framework for MEP components based on BIM and IoT using machine learning algorithms. *Automation in Construction*, **112**, 103087, 2020. DOI: 10.1016/j.autcon.2020.103087.

[22] Jardine, A.K.S., Lin, D. & Banjevic, D., A review on machinery diagnostics and prognostics implementing condition-based maintenance. *Mechanical Systems and Signal Processing*, **20**, pp. 1483–1510, 2006. DOI: 10.1016/j.ymssp.2005.09.012.

[23] UNI EN ISO 19650 Standard Organization and digitization of information about buildings and civil engineering works, including building information modelling (BIM): Information management using building information modelling. Part 3: Operational phase of the assets.

[24] Telikani, A., Gandomi, A.H. & Shahbahrami, A., A survey of evolutionary computation for association rule mining. *Information Sciences*, **524**, pp. 318–352, 2020. DOI: 10.1016/j.ins.2020.02.073.

[25] Salamak, M., Jasiński, M., Płaszczyk, T. & Żarski, M., Analytical modelling in Dynamo. *Transactions of Vsb*, Technical University of Ostrava, Civil Engineering Series, **18**(2), pp. 36–43, 2018. DOI: 10.31490/tces-2018-0014.

INTEGRATED SCALABLE FRAMEWORK FOR SMART ENERGY MANAGEMENT

AHMED AL-ADAILEH & SOUHEIL KHADDAJ
Science, Engineering and Computing Faculty, Kingston University – London, UK

ABSTRACT

The planet's resources experience fundamental troubles and unjust utilization. A large portion of the destruction is set off by using the planet's resources to produce energy of all kinds. To help with reverting the situation, there are mainly two approaches: firstly, to consider generating energy from clean and renewable resources, and secondly, to reduce the consumed energy by applying energy management systems. Due to the high energy consumption within the household sector, this paper aims to propose a dedicated household framework that tracks, predicts and manipulates the energy consumption of almost all appliances in the household; a sample household appliance is used to illustrate the main approach. The system is capable to track energy consumption and other related data directly from smart household appliances using their native connectivity and application programming interface capabilities, or from conventional appliances after equipping them with appropriate sensors and various Internet of Things (IoT) hardware. Once enough energy data is gathered, machine learning technologies will be applied to enhance the dataset and establish a solid background to predict energy consumption and apply the most suitable strategy from the available three strategies which most fits the appliance category. A case study implemented on a sample household appliance shows a possibility of reducing energy consumption by up to 22% by making a decision to replace the appliance with a more efficient one.

Keywords: household energy, household sector, IoT, predictive analysis, machine learning.

1 INTRODUCTION

Energy management systems can be defined as a set of rules and procedures applied to manage the energy journey beginning from production, through transfer and ending to consumption, to achieve the highest efficiency levels [1]. This paper will focus on proposing a framework to manage energy in the household sector because household energy consumption is considered a big portion of the total consumed energy worldwide, which is over 20,000 trillion Btu [2]. However, applying an energy management system in the household sector is a very challenging mission [3] and comes with many restrictions which can be summarized as: firstly, the challenge of reducing energy consumption while keep meeting the same level of comfort, which enhances the motivation of household occupants to continuously apply and use the system. Secondly, most household appliances are conventional appliances that lack connectivity capabilities. Thirdly, this field still does not offer a set of standards. Fourthly, the costs of the related technologies are still high and lack high levels of accuracy. Fifthly, there are still not enough reliable and dynamic open-source platforms that offer basic and common components which could be used for further adjustments and developments to gain more speed, reliability and higher quality.

Several approaches can be found in the literature, which can be grouped into four main groups: firstly, high-level energy management systems, which provide an abstract level of how to save energy without going into details. An example is the "Energy Management System designed for real Low-Voltage Distribution Networks with a High Penetration of Renewable Energy Sources" [4]. Secondly, the low-level systems, suggest building components on embedded systems and offer various interfaces. The main advantage of this approach is taking that of the host systems, and offering a high degree of flexibility to develop

WIT Transactions on Ecology and the Environment, Vol 255, © 2022 WIT Press
www.witpress.com, ISSN 1743-3541 (on-line)
doi:10.2495/EPM220121

new and adjust current components. A good example of them is the Open Gate for Energy Management Alliance (OGEMA) 2.0 [5], illustrated in Fig. 1, which offers several basic and deeply integrated components related to security, authentication and user management modules.

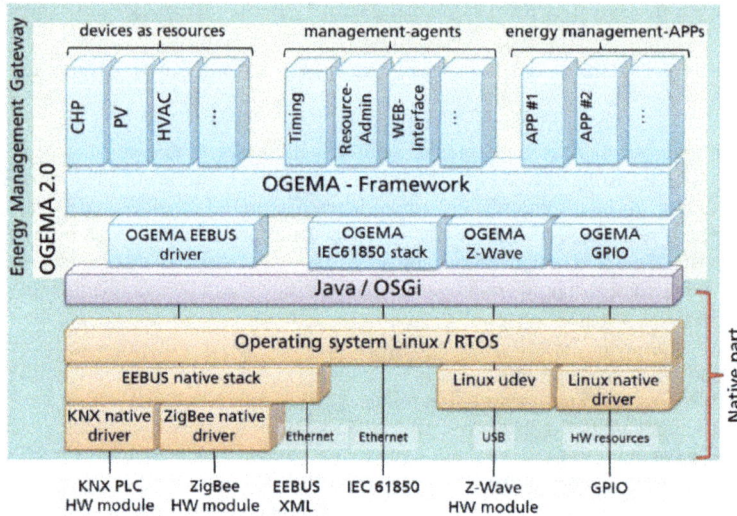

Figure 1: Energy management gateway architecture using OGEMA 2.0 [5].

Thirdly, the smart home energy management systems, which focus on utilizing an energy management system assuming having smart appliances, by providing advanced visualisations and illustration components, controlling possibilities including scheduling and routines to optimize the consumption, however, they did not provide a clear approach on how to deal with legacy and conventional devices. Self-Scheduling Model for Home Energy Management [6], and Smart Home Energy Management Framework dedicated to Internet of Things (IoT) Networks [7] are examples of these.

Fourthly, as illustrated by frameworks proposed by Bhayo et al. [8], [9] and Sutikno et al. [10], the integrated energy management systems supported by various hybrid clean energy generation schemes are equipped with storage capabilities to enhance sustainability and support the environmentally-friendly approach. However, the missing of using data-driven and prediction techniques to be able to deal with the near future events to act accordingly is considered a drawback which may reduce the efficiency of these types of frameworks. Also, relying on not fully-developed battery-based storage techniques may negatively affect the overall system reliability and dependability.

Finally, the data-driven predictive analysis-based systems, make use of machine learning (ML), artificial intelligence and IoT technologies. This approach can be seen in many frameworks that suggest using different predictive algorithms to provide energy consumption forecasting, such as the data-driven distributionally robust optimization based framework proposed by Saberi et al. [11], and the framework proposed by Pinto et al. [12] that makes use of the deep reinforcement learning approach..

Important to mention that these approaches suffer from the missing integrability of data retrieved from various hardware belonging to various vendors. Since this field suffers from missing standards, there is an essential need to offer a high level of integrability and data bridging to allow using different hardware, different standards and different data structures under one umbrella.

The next section describes the proposed framework which is an attempt to overcome most of the previously mentioned drawbacks and challenges.

2 THE PROPOSED FRAMEWORK

The proposed framework, illustrated in Fig. 2, is built upon the fact that household devices can be divided into three main categories: (1) Run-on-demand devices such as lights, (2) should-not-be-interrupted devices, that should not be manually switched on/off such as the refrigerator, (3) schedulable devices, these are devices that can be interfered with externally by switching them on/off; an example of these is the under-sink water heater or heating ventilation air conditioning systems.

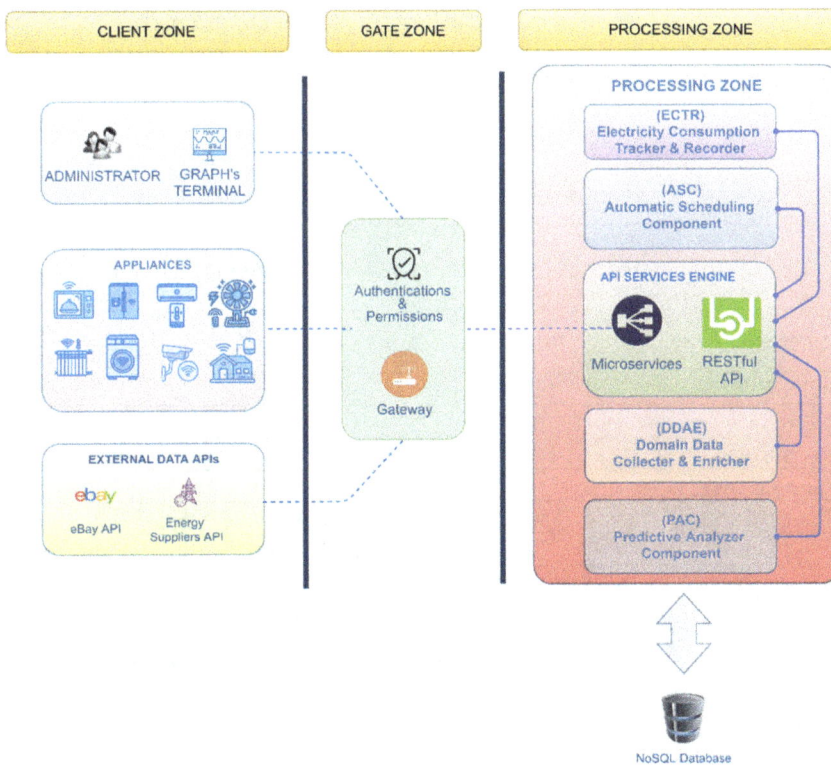

Figure 2: The proposed "Integrated Scalable Framework for Smart Energy Management".

It also suggests three different strategies to track, forecast and control the energy consumption of each group of devices accordingly. These strategies are: (1) Substituting devices based on energy consumption, which suggests tracking the energy consumption of the device, then suggests substituting it with a more energy-efficient one. Important to mention that old, inefficient appliances may only be used as spare parts, or to substitute other similar devices with higher energy consumption. (2) Substituting devices based on the usage percentage: this approach utilizes related artificial intelligence technologies to decide whether the device's size and volume are suitable for the targeted usage, for example, suggest substituting a 12 kg washing machine with a smaller one in a single-person household. (3) Automatic scheduling, which concentrates on tracking the energy consumption and the usage periods of a device, then predicting the future usage, and adjusting the running periods accordingly.

This paper focuses on the second strategy. Not each strategy applies to any household appliance, Table 1 shows which strategy can be applied to which device's group, whereas Table 2 shows a list of mainly used appliances, and their wattage, energy consumption (kWh) and average daily running periods in the household where the framework is applied.

Table 1: Device's groups and correspondent strategies.

Strategies	Device groups		
	Should not be switched off devices	Scheduled switch off/on devices	Run-on demand devices
Substituting devices based on energy consumption	X	X	X
Substituting devices based on the usage percentage	X	X	X
Automatic scheduling	–	X	–

Table 2: List of main appliances exist in the household where the case study takes place.

Appliances	Wattage	Energy consumption per hour	Average daily running periods
56 inch LED TV	125 W	0.13 kWh	3 hours
Refrigerator	300 W	0.3 kWh	24 hours
Coffee maker	800 W	0.8 kWh	0.5 hour
Vacuum cleaner	700 W	0.7 kWh	1 hour
Air fryer	1500 W	1.5 kWh	0.2 hour
Amazon Echo	3 W	0.003 kWh	3 hours
Bread toaster	800 W	0.8 kWh	0.3 hour
Clothes dryer	3000 W	3 kWh	0.4 hour
Laptop	250 W	0.3 kWh	4 hours
Electric iron	1200 W	1.1 kWh	0.3 hour
Under-sink water heater	1800 W	0.55 kWh	6 hours

As seen in Fig. 2, the framework consists of three main zones: (1) Client Zone: which consists of three main components including administrators, and a terminal to show the related graphs. Also, it includes the farm of related household appliances equipped with sensors. Finally, it has the external data application programming interfaces (APIs) to retrieve additional related external data. (2) Gate Zone: which is considered as a bridge between Client Zone and Processing Zone. It is responsible for passing the traffic and communication, applying the necessary authentication and permission procedures. (3) Processing Zone: Depending on the number of involved households, this zone may physically reside in the household, or the cloud. It consists of all components needed to process the data retrieved from the Client Zone which includes:

(a) Electricity consumption tracker and recorder which is responsible for tracking, cleansing, preparing and saving the energy consumption for each device.
(b) The automatic scheduling component is responsible for collecting all related data, then applying the ML techniques to predict the future behaviours of the household's occupants and devices, adjust their running periods, and offer the same comfort level, and reduce the energy consumption.
(c) API service engine (ASE) this component is the central point where all requests come in. It uses the state-of-art microservices topology to support the scalability to deal with an unlimited number of households.
(d) Domain data collector and enricher: the prediction process requires a huge number of data collected from the surrounding environment, such as internal and external temperatures, holiday periods, traffic, etc.; this component is responsible for choosing the suitable data nodes, the collecting and preparing them for example by applying some anomaly detection steps, and applying feature engineering aspects. Finally
(e) The predictive analyzer component where the ML process takes place by applying the cross-industry standard process for data mining beginning with business understanding, then data understanding, followed by data preparation, modelling, evaluation and deployment.

All components are connected to a NoSQL Database (MongoDB) which is chosen due to performance and scalability reasons. Data objects are collected from different sources with different structures so there is a need to be able to save these objects as they arrive without a need to a fix pre-structure.

The aim of applying ML and prediction techniques is to enhance the overall dataset giving a solid background to make proper decisions when applying the substitution strategy. Moreover, using the proposed framework on a wider scale with a huge number of households may deliver other advantages than directly reducing the energy consumption; these can be summarized as: (1) Providing the local energy providers with accurate consumption levels, so they can adjust their energy production accordingly. (2) Offering the appliances' manufacturers real-life running operational parameters of their devices, to let them spot the potential weak points and fix them in the upcoming versions. (3) Collected anonymous data may also assist the relevant governmental agencies to evaluate the reality and designing appropriate laws and regulations. Once enough dataset is collected from historical and predicted sources, the following equations will be applied to calculate the amount of saved energy:

$$\text{Current Appliance Daily Average Consumption}_{(kWh)}(CADAC)$$
$$= MPC_{(kWh)} * ANHRRD_{(hr)}$$

where:
MPC = measured and predicted consumption in kWh; and ANHRRD = average number of hours refrigerator runs daily.

$$\text{New Appliance Daily Average Consumption}_{(kWh)}(NADAC)$$
$$= NAC_{(kWh)} * ANHRRD_{(hr)}$$

where NAC = new appliance consumption in kWh.

3 CASE STUDY
As mentioned before, due to its integrability and scalability nature, the proposed framework establishes a wide basis that supports several strategies to reduce energy consumption in the

household. However, in this case study, the focus will be on the first device's group "Should-not-be-switched-off Devices" such as refrigerator and the first strategy "Substituting Devices based on energy consumption".

The implementation phase aims to predict the energy consumption of the refrigerator, so it is possible to enhance the overall dataset which consists of the tracked energy consumption data and the predicted ones. This enables an accurate decision to replace or keep the refrigerator. It begins with collecting a number of data nodes from the surrounding area where the refrigerator is operating.

The data will be collected from three main sources: (1) Constants related to the device such as type, and size, which are entered by system administrators; (2) Sensing data retrieved from sensors; and (3) Specially constructed systems that use artificial intelligence techniques to read settings from conventional devices, such as the Refrigerator Temperature Settings Panel Reader shown in Figs 3 and 4. Finally, (4) External data retrieved from various APIs such as traffic and e-commercial platforms.

RECHS Transformers - Refrigerator Temperature Settings Reader

Using a specially installed camera, on daily intervals, photos from the refrigerator temperature setting panel is taken and sent to the system to guess the temperature level. Dataset consists of following variables:
- Taken images
- Date/Time
- Guessed Temperature
- Accuracy

Images	Date / Time	Guessed Temperature	Accuracy
	4/7/2020 14:33:18	level-2	high
	11/7/2020 14:33:18	level-2	high
	3/7/2020 14:16:01	level-3	high

Figure 3: Transformers: Refrigerator temperature settings panel reader component.

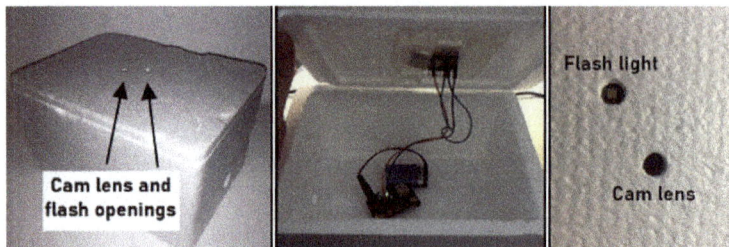

Figure 4:   Refrigerator camera module used to feed the frig temperature settings panel transformer with proper images.

Then a list of verification, and cleaning actions are taken to ensure that only relevant data is considered. The outcome is shown in Table 3.

Table 3: Initial variables, sources, database field type, format, possible data range and some examples, with applied pre-processing.

Variable name	Source	Field type/format	Example data/values	Action	Justification
Datetime	System	Datetime	2021-03-10 13:59:45	Removed	Not relevant for regression algorithms predictions
Internal temperature	Sensor	Float		Accepted	Shows high correlation with the target-feature
External temperature 5 cm	API	Float		Removed	Low target-correlation
External temperature 5 m	API	Float		Removed	See "External temperature 5 cm"
External temperature measured	Sensor	Float		Removed	See "External temperature 5 cm"
External relative humidity	API	Integer	0–100%	Removed	Low correlation score
External relative humidity measured	API	Integer	0–100%	Removed	Removed due to low correlation score
Internal relative humidity measured	Sensor	Integer	0–100%	Removed	A high percentage of missing data
Weather condition	API	Integer	Rainy, windy, stormy	Removed	Low variance score (0.24877371)
Refrigerator fullness	ML	Integer	0–100%	Accepted	It shows a high correlation score with the target-dependent feature
Occupants	ML	Integer		Accepted	The correlation score is not high, but also not low enough to discard this feature
Refrigerator temperature	ML	Integer	Level 1 – 6	Accepted	A high correlation score was calculated
Energy consumption	Sensor	Float	Watts/hour	Accepted	This is the target-dependent feature to be predicted measured in watts
Times door opened	Sensor	Integer		Accepted	Despite high feature-wise correlation with "Duration door left opened", the field is kept to get better predictions
Duration door left opened	Sensor	Integer		Accepted	See "Times door opened"
Day type	System	Varchar	Weekend, weekday	Altered	Altered to (weekend: 0,1) because the effect of the day type affects the number of occupants staying at the household. The day type itself does not affect the prediction

Part of the data preparation is detecting the anomalies. This is done for the accepted datasets to make sure no anomalies are existing which may negatively be affecting the quality of the predicted models. Fig. 5 illustrates this process done for some selected data nodes.

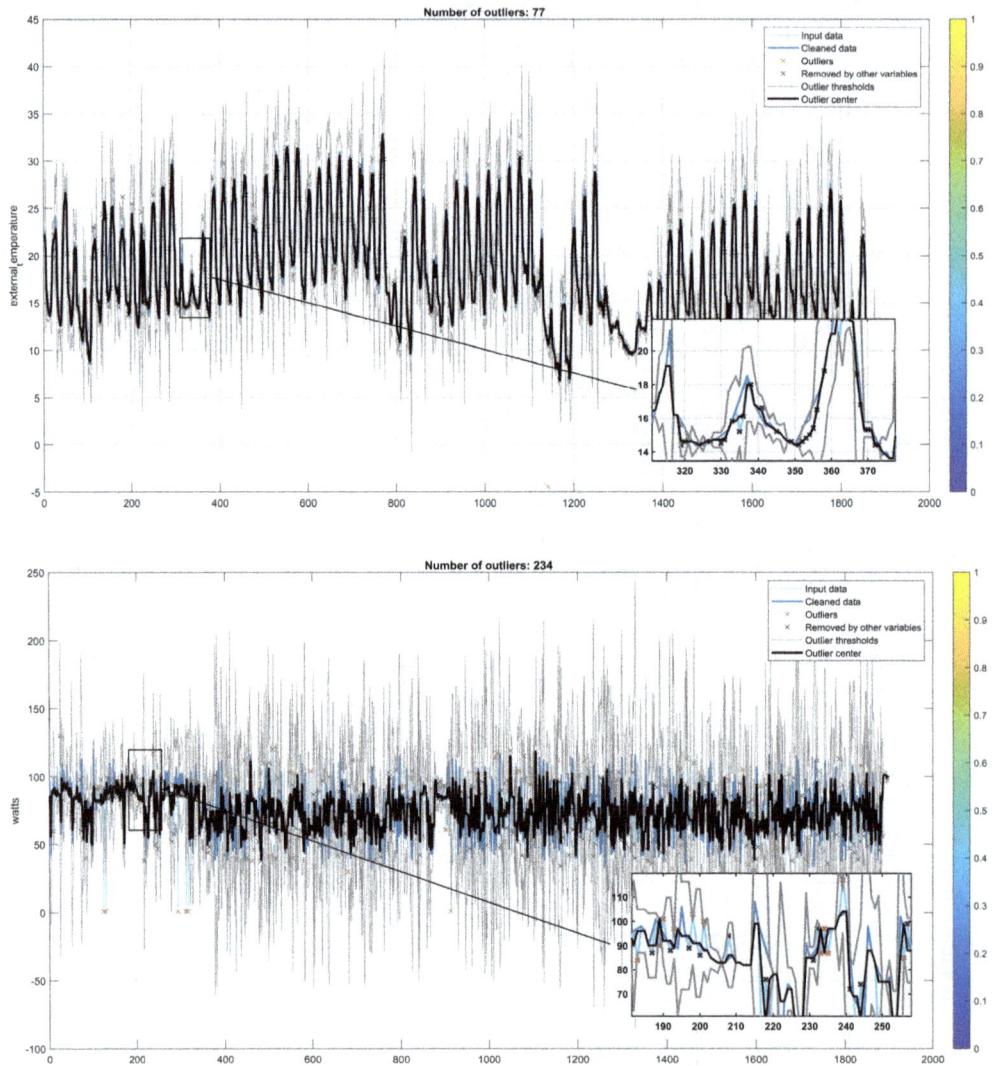

Figure 5: Anomaly detection plots of some selected features.

Once data is prepared, the modelling phase can begin to select the most suitable algorithm. Table 4 shows the resulted metrics which indicates that the models created using the Linear Regression (Multiple) are the most suitable for this purpose, where Fig. 6 shows the comparison among various algorithms and the targeted wattage derived from the verification dataset.

Table 4: Calculated regression algorithms' metrics.

Model	MSE	RMSE	MAE	R2
Polynomial regression	215.542	14.681	9.871	0.397
Linear regression (multiple)	430.948	20.759	15.519	−0.205
Random forest	8.363	2.892	1.708	0.977
Tree	8.332	2.886	1.685	0.977
Linear regression (simple)	22.513	4.745	3.359	0.937
kNN	13.519	3.677	2.254	0.962

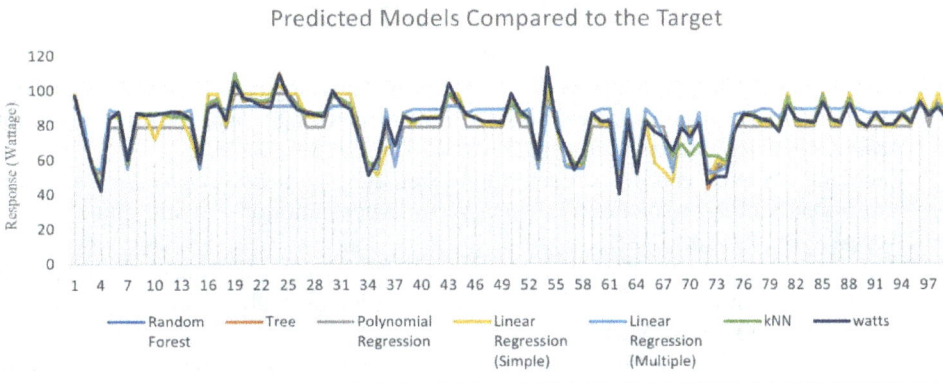

Figure 6: Regression prediction models compared to the verification wattage.

The collected historical data and the predicted data are used to save energy consumption by applying the previously mentioned substation strategy using the eBay product search API which returns a list of refrigerators with similar features to the sample refrigerator but have better energy efficiency. The saving of each result can be calculated using the previously mentioned equations, as follows:

$$\text{Energy Saving} = \frac{\left(\text{CADAC}_{(kWh)} * 365_{(day)}\right) - \left(\text{NADAC} * 365_{(day)}\right)}{\text{CADAC}_{(kWh)} * 365_{(day)}} * 100\%$$

Applying this equation to different refrigerators shows that a refrigerator with the energy efficiency class A+++ may save energy up to 22%, as follows:

$$\text{CADAC}_{(kWh)} = 0.081 * 8 = 0.648 \text{ kWh}$$

$$\text{NADAC}_{(kWh)} = 0.063 * 8 = 0.504 \text{ kWh}$$

$$\text{Energy Saving} = \frac{(0.648 * 365) - (0.504 * 365)}{0.648 * 365} * 100\% = 22.2\%$$

This result can be even more enhanced when applying the framework to other appliances that can be switched on/off such as the under-sink immersion water heater. For this case, a manual or automatic switch mechanism, can be applied to cut off the energy consumption when the device is not used by setting a date, time and interval to start/stop the appliance.

4 CONCLUSION

This paper proposed a data-driven, predictive, scalable and integrated framework, which is applied to a household with a combination of smart and conventional appliances to gather various data nodes to predict the energy consumption for a sample appliance, using various regression algorithms which were evaluated via several evaluation metrics to decide for the most suitable model. Applying the mentioned strategy has shown an energy reduction of up to 22% for a sample device, the refrigerator, from the uninterruptible appliances category. The future work may include implementing the framework on a wider scale, covering more households, more appliances and applying other strategies.

REFERENCES

[1]   Ganesh, A.H. & Xu, B., A review of reinforcement learning based energy management systems for electrified powertrains: Progress, challenge, and potential solution. *Renewable and Sustainable Energy Reviews*, **154**, 111833, 2022.

[2]   Enerdata, World energy consumption statistics. *Yearbook.enerdata.net*. 2022. https://yearbook.enerdata.net/total-energy/world-consumption-statistics.html. Accessed on: 27 Apr. 2022.

[3]   Wang, X., Mao, X. & Khodaei, H., A multi-objective home energy management system based on Internet of Things and optimization algorithms. *Journal of Building Engineering*, **33**, 101603, 2021.

[4]   Kelm, P. et al., Hardware-in-the-loop validation of an energy management system for LV distribution networks with renewable energy sources. *Energies*, **15**, 2561, 2022.

[5]   Nölle, C., OGEMA. https://www.ogema.org. Accessed on: 29 Apr. 2022.

[6]   Javadi, M.S. et al., Self-scheduling model for home energy management systems considering the end-users discomfort index within price-based demand response programs. *Sustainable Cities and Society*, **68**, 102792, 2021.

[7]   Aliero, M.S., Qureshi, K.N., Pasha, M.F. & Jeon, G., Smart home energy management systems in Internet of Things networks for green cities demands and services. *Environmental Technology and Innovation*, **22**, 101443, 2021.

[8]   Bhayo, B.A., Al-Kayiem, H.H., Gilani, S.I.U. & Ismail, F.B., Power management optimization of hybrid solar photovoltaic-battery integrated with pumped-hydro-storage system for standalone electricity generation. *Energy Conversion and Management*, **215**, 112942, 2020.

[9]   Bhayo, B.A., Al-Kayiem, H.H., Gilani, S.I.U., Khan, N. & Kumar, D., Energy management strategy of hybrid solar-hydro system with various probabilities of power supply loss. *Solar Energy*, **233**, pp. 230–245, 2022.

[10]   Sutikno, T., Arsadiando, W., Wangsupphaphol, A., Yudhana, A. & Facta, M., A review of recent advances on hybrid energy storage system for solar photovoltaics power generation. *IEEE Access*, **10**, pp. 42346–42364, 2022.

[11]   Saberi, H., Zhang, C. & Dong, Z.Y., Data-driven distributionally robust hierarchical coordination for home energy management. *IEEE Transactions on Smart Grid*, **12**(5), pp. 4090–4101, 2021.

[12]   Pinto, G., Deletto, D. & Capozzoli, A., Data-driven district energy management with surrogate models and deep reinforcement learning. *Applied Energy*, **304**, 117642, 2021.

Author index

WIT*PRESS* ...for scientists by scientists

Energy Resources and Policies for Sustainability

Edited by: A. TADEU, University of Coimbra, Portugal

An increasing interest in renewable energy resources and the search for maintainable energy policies have inspired the research contributions included in this book.

Energy production and distribution need to respond to the modern world's dependence on conventional fuels. To achieve this, collaborative research is required between multiple disciplines, including materials, energy networks, new energy resources, storage solutions, waste to energy systems, smart grids and many other related subjects.

Energy policies and management are of primary importance for sustainability and need to be consistent with recent advances in energy production and distribution. Challenges lie as much in the conversion from renewable energies such as wind and solar to useful forms like electricity, heat and fuel at an acceptable cost (including environmental damage) as in the integration of these resources into existing infrastructure.

ISBN: 978-1-78466-371-1   eISBN: 978-1-78466-372-8
Published 2020 / 320pp

www.ingramcontent.com/pod-product-compliance
Lightning Source LLC
Chambersburg PA
CBHW062008190326
41458CB00009B/3008